# THE MOLECULES OF LIFE

*Readings from*

**SCIENTIFIC
AMERICAN**

# THE MOLECULES
# OF LIFE

W. H. Freeman and Company
*New York*

THE COVER

The cover displays an end-on view of the DNA double helix, the molecule that encodes genetic information and has become the emblem of molecular biology, the theme of this book. The computer-generated image offers a wide-angle look along the axis of the $B$ form of the double helix. The sugar and phosphate groups comprising the backbone of one of the two strands of the molecule are diagrammed in red, the elements of the other backbone in green. The strands are linked by paired bases: a purine (*blue*) on one strand pairs with a pyrimidine (*pink*) on the other strand. The swirling cloud of dots is the solvent-accessible surface: the outermost surface of the DNA molecule, defined by individual atoms, with which other molecules interact. Arthur J. Olson of the Research Institute of Scripps Clinic generated the image. He worked with the computer-graphics language GRAMPS (which he developed with T. J. O'Donnell), a molecular-modeling package (developed with Michael L. Connolly) called GRANNY and MS, a program for calculating dot surfaces, written by Connolly.

Library of Congress Cataloging in Publication Data

Main entry under title:

The Molecules of Life.

    "Readings from Scientific American."
    "The eleven chapters in this book originally appeared as articles in the October 1985 issue of Scientific American."—T.p. verso.
    Bibliography: p.
    Includes index.
    1. Molecular biology—Addresses, essays, lectures.
I. Scientific American.
QH506.M665 1986       574.8'8       85-27598
ISBN 0-7167-1792-1
ISBN 0-7167-1783-2 (pbk.)

The eleven chapters in this book originally appeared as articles in the October 1985 issue of SCIENTIFIC AMERICAN.

Printed in the United States of America

1 2 3 4 5 6 7 8 9 0  KP  4 3 2 1 0 8 9 8 7 6

# CONTENTS

FOREWORD    *vii*

**1  THE MOLECULES OF LIFE,** by Robert A. Weinberg    1
*Presenting a book on the powerful techniques·and remarkable findings of the new molecular biology.*

**2  DNA,** by Gary Felsenfeld    13
*The double helix can change its shape, enabling it to interact with various regulatory molecules.*

**3  RNA,** by James E. Darnell, Jr.    25
*Now it translates DNA into proteins, but it may itself have been the very first genetic material.*

**4  PROTEINS,** by Russell F. Doolittle    37
*Genes encode proteins; proteins in turn, by means of selective binding, do almost everything else.*

**5  THE MOLECULES OF THE CELL MEMBRANE,**    49
by Mark S. Bretscher    *A bilayer of lipids, in which proteins are embedded, controls traffic into and out of the living cell.*

**6  THE MOLECULES OF THE CELL MATRIX,**    59
by Klaus Weber and Mary Osborn    *The framework of varied proteins that gives form to the cell is being analyzed by new techniques.*

**7  THE MOLECULES OF THE IMMUNE SYSTEM,**    71
by Susumu Tonegawa    *An almost infinitely diverse battery of proteins recognize foreign invaders and defend against them.*

**8  THE MOLECULAR BASIS OF COMMUNICATION BETWEEN CELLS,** by Solomon H. Synder    *Hormones and neurotransmitters seem very different, but some molecules act as both.*    83

**9  THE MOLECULAR BASIS OF COMMUNICATION WITHIN THE CELL,** by Michael J. Berridge    *A few "second messengers" relay signals regulating a wide variety of cellular responses.*    95

**10 THE MOLECULAR BASIS OF DEVELOPMENT,** 107
by Walter J. Gehring     *A bit of DNA called the homeobox helps to orchestrate development in a startling array of animals.*

**11 THE MOLECULAR BASIS OF EVOLUTION,** 119
by Allan C. Wilson     *By tracking mutations in DNA, molecular biologists gain new insights into organismal evolution.*

THE AUTHORS     132

BIBLIOGRAPHIES     134

INDEX     136

# FOREWORD

With molecular scissors called endonucleases, with invisible probes such as monoclonal antibodies and snippets of DNA, and with a host of other powerful tools, biologists have engineered the major scientific revolution of our era. They have learned to understand, manipulate, and transform the very stuff of life: the nucleic acids and proteins that are the elemental (but far from elementary) components of living things. The new molecular biology has implications far beyond the laboratory. To deal with them, the human species will need to gain ethical and moral maturity. A good starting point is an appreciation of the diversity, intricacy, and beauty this robust science presents to our understanding.

Biologists have dealt with molecules for 150 years or so, but perhaps the molecular era of biology can be said to have had its subtle and unnoticed beginning in 1944. Oswald Avery, Colin MacLeod, and Maclyn McCarty then identified the "transforming principle," a factor mediating inherited change in the pneumococcus, as a particular molecule: something called deoxyribonucleic acid. In other words, the genetic material is DNA. The news did not immediately transform biochemistry or genetics. That began after 1953, when the three-dimensional structure of DNA was resolved by James Watson and Francis Crick. In the ensuing years, the genetic code and the flow of information from DNA to proteins were clarified; the molecular basis of cell structure, the immune response, neural and hormonal signalling, and embryonic development began to be understood.

Within the past 10 years, the powerful new techniques for manipulating DNA have transformed the investigation not only of genes and their protein products, but also of a wide range of cellular structures and mechanisms. The central maneuver is the cloning of genes. The significance of cloning and other techniques and the broad implications of the new molecular biology are described in the first chapter.

The next three chapters deal with three major polymers of life: DNA, RNA, and proteins. Genetic information is stored and replicated as DNA, interpreted by RNA, and ultimately expressed as proteins. Expression is controlled primarily by the selection of particular stretches of DNA for transcription into messenger RNA. The DNA of genes is packaged with protein, coiled and supercoiled in a form called chromatin. The packaging is undone at one site or another, and then various protein molecules interact with particular nooks and crannies of the DNA to initiate transcription. The multiplicity and complexity of these interactions make it clear that the conformation of the double helix is almost as variable as DNA's linear sequences. Once transcribed into messenger RNA, the information specified by a gene must still undergo processing before it can be translated into protein. Evolutionary studies and what has recently been learned about the processing provoke speculation that RNA may have been the first genetic material.

Proteins are the primary products of genes; all other molecules—including sugars and fats and DNA itself—are the products of biochemistry conducted by the proteins called enzymes. Whether as enzymes, structural components, or the messengers and receptors of communications systems, proteins function by virtue of their complex shapes, which fit snugly with other molecules.

The molecules of particular cellular structures and systems are described in the next five chapters. The living cell is both set apart from and linked to other cells by a delicate but authoritative wall, the cell membrane, whose mobile elements maintain the integrity of the cell while at the same time admitting nutrients and a host of other molecules into the cell's cytoplasm. The cytoplasm and its metabolism are organized by a cell matrix: a skeletal framework whose composition and structure have been defined in detail by investigators working with fluorescent-labeled monoclonal antibodies. These proteins are highly specific laboratory versions of the antibodies a vertebrate animal deploys in defense against certain foreign invaders. Antibodies are but one component of the immune system; others are cell-surface receptors whose molecular structure, only now being deciphered, shows they are close cousins of the antibodies.

The individual cells of a multicellular organism need to communicate with one another. At the molecular level the two communciation networks, the nervous system and the endocrine system, turn out to be rather similar; they even depend on many of the same messenger molcules. The arrival of such messengers at the surface of a target cell is detected by specialized receptors. The detection triggers a series of reactions in the cell membrane that activate an internal messenger; it is this second messenger that prompts the cell to do what the external messenger ordained: to divide, say, or to change shape, or to synthesize and secrete a particular substance.

The last two chapters deal with the fresh contribution of molecular biology to two very different kinds of history: the embryonic development of an individual organism and the evolution of individual species. Cloning has been a major tool in the recent identification of several master genes that control the timing of developmental events and the spatial organization of the embryo. Evolution results from changes in genes, the result of mutation and natural selection. The molecular biologist's ability to directly compare the proteins and DNA of different species has provided a new quantitative measure of evolution.

The new biology described in this book (whose chapters first appeared as articles in the October, 1985, single-topic issue of *Scientific American*) is well characterized by Robert Weinberg in the first chapter: "The beauty and wonder of nature are nowhere more manifest than in the submicroscopic plan of life" the reader is about to encounter.

THE EDITORS*

*October 1985*

*PRESIDENT AND EDITOR: Jonathan Piel

BOARD OF EDITORS: Armand Schwab, Jr. (Associate Editor), Timothy Appenzeller, John M. Benditt, Peter G. Brown, David L. Cooke, Jr., Ari W. Epstein, Michael Feirtag, Robert Kunzig, Philip Morrison (Book Editor), James T. Rogers, Joseph Wisnovsky.

# 1

# THE MOLECULES OF LIFE

# The Molecules of Life

*Introducing a volume about the new biology, which seeks to explain the molecular mechanisms underlying biological complexity. It has given rise to an industry, and to new ways of thinking about life*

by Robert A. Weinberg

Biology in 1985 is dramatically different from its antecedents only 10 years ago. New investigative techniques have made commonplace many experiments that were previously far beyond the reach of even the cleverest experimental biologist. The new molecular biology has done much more than expand the repertoire of laboratory techniques. It has, with remarkable rapidity, established a biotechnology industry. More important, it has changed the ways people think about living things, from bacteria to human beings.

Biology has traditionally been a descriptive science. The multitude of living organisms were catalogued, their traits enumerated and their structures examined on a gross or a microscopic level. In thus describing organismic traits, or phenotypes, biologists confronted only the consequences of biological processes, not the causative forces. An experimenter could watch a muscle contract or an embryo develop, but such observation alone could not provide the clues that were needed for any real understanding of underlying mechanisms.

The ability to observe was greatly extended by the development of microscopic techniques that made it possible to visualize cells and subcellular organelles. Electron microscopy pushed the limits of visualization even further: the fine structure of cells could be resolved with great precision. This advance led to the uncovering of still more structures and phenomena whose causative mechanisms remained unexplained. The explanations clearly lay with elements even smaller than the cellular components observed by microscopists.

It became apparent that the ultimate casual mechanisms behind many biological phenomena depend on the functioning of specific molecules inside and outside the cell. This *Scientific American* book describes how investigators think about biological systems in terms of their molecular components. The chapters that follow are permeated by the assumption that to describe biological phenomena is far less interesting than to elucidate the molecular mechanisms underlying them. The molecular biologists who present their work here manipulate things they will never see. Yet they work with a certainty that the invisible, submicroscopic agents they study can explain, at one essential level, the complexity of life.

The newly gained ability to describe and manipulate molecules means the biologist is no longer confined to studying life as the end product of two billion and more years of evolution. The new technology has made it possible to change critical elements of the biological blueprint at will, and in so doing to create versions of life that were never anticipated by natural evolution. In the long run this may prove to be the most radical change deriving from the power to manipulate biological molecules.

Among the many kinds of biological molecules in the living cell, three have attracted the greatest attention: protein, RNA and DNA. They are macromolecules, large molecules that are linear polymers built up from simple subunits, or monomers. It was the proteins that attracted the lion's share of attention until 20 years ago. The reason, in retrospect, is clear. Certain specialized tissues accumulate large amounts of only one kind of protein. Red blood cells have almost pure hemoglobin, cartilage consists largely of collagen and hair is largely keratin. Biochemists studied such proteins first because they could be isolated in pure form, purity being a prerequisite to further study.

As an array of sophisticated biochemical techniques emerged it became possible also to purify those proteins found only in trace amounts within the complex chemical soup of a living cell. Biochemists could now concentrate on proteins that function as enzymes, catalyzing the several thousand biochemical reactions that in the aggregate constitute the metabolism of living cells. This work went well, because many of the reactions could be easily reconstructed in a test tube containing the proper mixture of reactants and catalyzing enzymes.

Yet in the past quarter of a century proteins have been gradually upstaged as objects of attention by the other macromolecules, first by RNA and more recently by DNA. There were two important reasons. The first one stems ironically from the great successes of protein biochemistry, which produced an avalanche of data on thousands of proteins and biochemical reactions. It soon became apparent that further study of individual trees gave little hope of understanding the entire forest. What was responsible for organizing and orchestrating this complex array of structures and processes? The answer lay not with the proteins

**DOUBLE HELIX OF DNA, the molecule that is the repository of genetic information and so may be considered the fundamental molecule of life, is seen from the side in the computer-generated image on the opposite page. The spheres represent individual atoms: oxygen is red, nitrogen is blue, carbon is green and phosphorus is yellow. Diagonal regions of the image delineate the sugar-phosphate backbone of the ladderlike helix; the horizontal elements, made up of nitrogen, carbon and oxygen atoms, are the base pairs that cross-link the two strands of the helix. The computer program eliminates the backbone on the far side of the structure. The image, which depicts the *B* form of DNA, was generated by the Computer Graphics Laboratory of the University of California at San Francisco.**

but with the study of genetics, and of the nucleic acids that carry genetic information.

The other reason nucleic acids, particularly DNA, have taken center stage is the advent of recombinant-DNA technology. In the course of the past decade biologists have learned to manipulate DNA in ways that (at least currently) are impossible for the protein chemist. DNA can be cut apart, modified and reassembled; it can be amplified to many copies; perhaps most telling, with DNA one can generate RNA and then protein molecules of wanted size and constitution. The central experimental maneuver in these manipulations is the cloning of genes, and it is cloning, more than any other single factor, that has changed the face of biology.

The groundwork for the cloning of genes was laid in 1953, when the

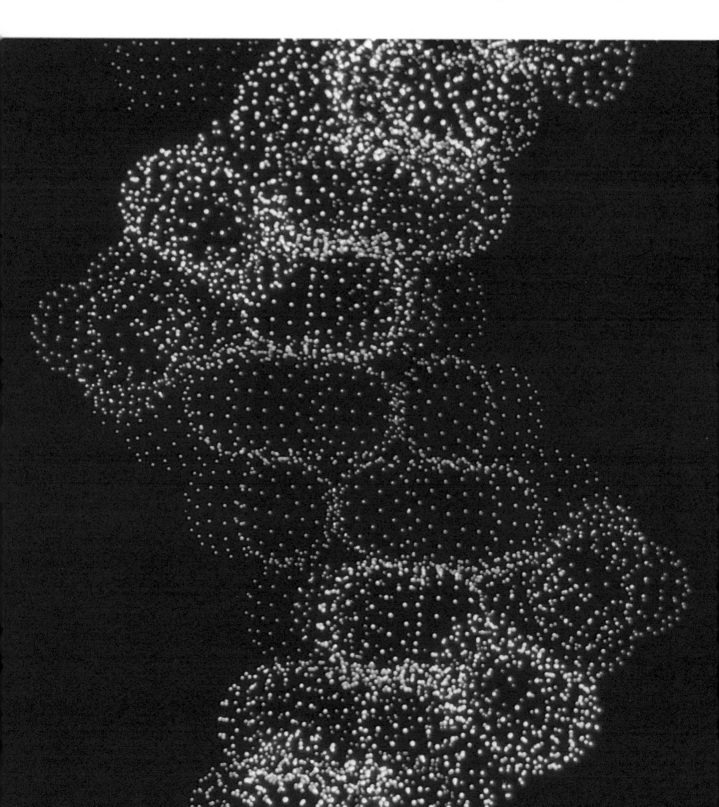

double-helical structure of DNA was perceived by James Watson and Francis Crick. A strand of DNA is a chain of nucleotides, each containing one of four bases: adenine (*A*), guanine (*G*), thymine (*T*) and cytosine (*C*). An *A* on one strand of the double helix pairs with a *T* on the other strand, and *G* pairs with *C*, so that the two strands are complementary. The sequence of bases specifies the order in which amino acids are assembled to form proteins. When the information in a gene is read out (expressed), its base sequence is copied (transcribed) into a strand of RNA. This messenger RNA (mRNA) serves as a template for the synthesis of protein: its base sequence is translated into the amino acid sequence of one protein or another.

The encoding of proteins is only a small part of DNA's function, and hence of its information content. To learn this and other simple facts it was necessary to first learn about the overall organization of DNA sequences and how the functional units of DNA—the individual genes—interact with one another in the total genetic repertoire of the organism, which is called its genome.

The genome of complex organisms resisted analysis until recently. Analysis of the gross biochemical properties of cellular DNA gave little hope of understanding the subtleties of genetic organization. The DNA content of even a bacterial cell is very large; the much larger genome of a mammalian cell carries some 2.5 billion base pairs of information arrayed along its chromosomal DNA. The base sequences are arranged in discrete compartments of information: the individual genes. There are between 50,000 and 100,000 genes in the genome of a mammal; each one is presumably responsible for specifying the structure of a particular gene product, usually a protein. Interest was therefore focused on studying individual genes, but that undertaking was doomed until recently by an inability to study single genes in isolation. In the absence of effective techniques of enrichment and isolation, individual cellular genes were abstractions. Their existence was suggested by genetic analysis, but their physical substance remained inaccessible to direct biochemical analysis.

A partial solution to this quandary came from the study of viruses. Their genome is very small compared with that of a cell and yet their genes are similar to those of the cells they infect. The DNA genome of one much studied animal virus, the SV40 virus of monkeys, has only 5,243 base pairs, in which are nested five genes. The analysis of an individual gene was therefore not confounded by a large excess of unrelated sequences. Moreover, the viral genome multiplies to several hundred thousand identical copies within an infected cell, and it was not hard to separate the viral DNA from the cellular DNA.

Once purified, the relatively simple viral DNA made it possible to study aspects of gene structure, the transcription and processing of messenger RNA and the synthesis of proteins that had previously been beyond reach. Still unresolved were the detailed structure and the base sequence of even the viral genome, whose 5,000-odd base pairs represented a daunting challenge to biochemists trained to take polymers apart one unit at a time. Then, in the mid-1970's, two revolutionary techniques became widely available that radically simplified the analysis of DNA structure.

The first of the techniques stemmed from the discovery of DNA-cleaving enzymes called restriction endonucleases. These enzymes, extracted from bacteria, cut a DNA molecule only at specific sequences that occur here and there along the DNA double helix. The much used endonuclease *Eco*RI, for example, cuts wherever it encounters the sequence *GAATTC*; *Sma*I cuts at *CCCGGG*, and so on. The sequences forming recognition sites occur with a certain statistical probability in any stretch of DNA.

Restriction enzymes have become powerful tools for experimenters. They establish convenient, fixed landmarks along the otherwise featureless terrain of the DNA molecule. They allow one to reduce a very long DNA molecule into a set of discrete fragments, each of them from several hundred to several thousand bases long. The fragments can be separated from one another on the basis of their size

**MOLECULAR ROSE WINDOW is a view along the axis of the *B* DNA double helix. In this image, also made by the Computer Graphics Laboratory, 10 consecutive nucleotide pairs along the helix are collapsed into a plane; the tenfold symmetry results from the fact that there are 10 component nucleotides per turn of the helix. The surfaces of the sugar and phosphate groups of the backbones are delineated by dots representing atoms: carbon (*green*), oxygen (*red*) and phosphorus (*yellow*). In the center the dots are absent, and so the skeletal structure of the bases, largely nitrogen (*blue*) and carbon, is left exposed.**

by gel electrophoresis. Each fragment can then be subjected individually to further analysis.

The other technical revolution had to do with the sequencing of DNA. Several procedures were invented by which the entire base sequence of a segment generated by restriction-enzyme cleavage can be determined with remarkable rapidity. These methods made it possible, for example, to establish the entire nucleotide sequence of the SV40 genome by 1978. Because the genetic code for translating a DNA sequence into an amino acid sequence was already known, the base sequences in certain regions of the genome could be translated into amino acid sequences. In this way the structure of SV40's proteins could be deduced from its DNA structure. Previously protein structure had been determined by the painstakingly slow biochemical analysis of individually isolated proteins; now the rapid sequencing of DNA could determine protein sequences in a fraction of the time. DNA sequencing also revealed other regions of the SV40 genome that are involved not in encoding proteins but in regulating the expression of genes and the replication of DNA.

Further progress depended on procedures for isolating individual cellular genes. Here success came from studies of bacterial genetics that were initiated in the early 1970's. The procedures for gene isolation that grew out of this work are based ultimately on the similarity of molecular organization in all organisms, from bacteria through mammals. As a result bacterial and mammalian DNA's are structurally compatible; DNA segments from one life form can readily be blended with the DNA of another form.

This similarity in DNA structure extends to many of the bacteriophages (viruses that infect bacteria) and plasmids (small DNA circles that parasitize bacteria). Phages inject their DNA into a bacterium, cause it to be replicated many times over, package the newly replicated DNA in viral-protein coats and kill the bacterium; the progeny phages released from the cell go on to infect other susceptible hosts. The plasmids are even simpler, and they have a more symbiotic relation with

DNA SUPERHELIX, in which the double helix is itself wound into a coil, is depicted in a space-filling computer model made by Nelson L. Max of the Lawrence Livermore National Laboratory. This is thought to be the form in which DNA is actually packed into chromosomes in the cell nucleus, with two turns of the superhelix being wound around a complex of histone proteins (not shown here). The model is based on coordinates determined by Joel L. Sussman and Edward N. Trifonov of the Weizmann Institute of Science.

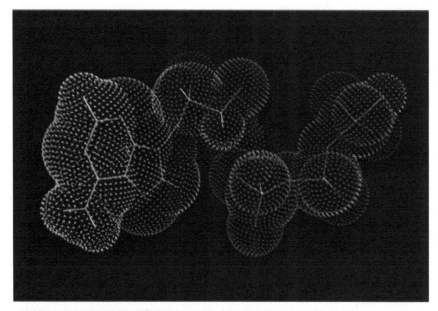

ATP (adenosine triphosphate) is the molecule that provides free energy for many biochemical reactions, including those required for the polymerization of DNA, RNA and protein. ATP is modeled in this image made by the Computer Graphics Laboratory. It is a nucleotide consisting of the base adenine (*left*), a ribose sugar and three phosphate groups (*right*). Energy is acquired when a third phosphate is added to adenosine diphosphate by the oxidation of fuel molecules or, in plants, by photosynthesis; energy is liberated when ATP is broken down, freeing this third phosphate group. The skeletal structure of the molecule is indicated by the lines; the dots delineate the effective surfaces of the constituent atoms.

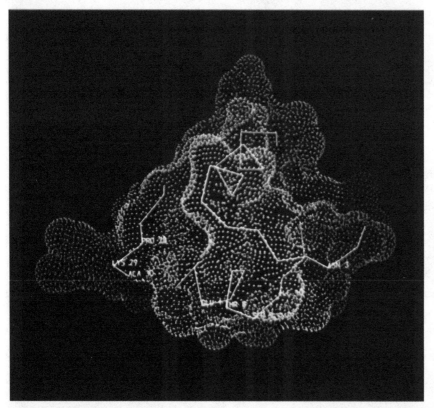

INSULIN MOLECULE, a hormone that has multiple functions, is depicted in a computer-generated model. It was developed by Elizabeth D. Getzoff, J. A. Tainer and Arthur J. Olson of the Research Institute of Scripps Clinic with the same software that generated the image on the cover of this book. Insulin is a small protein hormone made up of two short folded chains of amino acids. The lines trace the backbone of the two amino acid chains; the dots delineate the solvent-accessible surface. Coloring reflects the relative mobility of constituent atoms: the atoms shown in red and orange are the ones most subject to excursion from their mean position in a crystal of insulin and those in green and blue are the least mobile.

the bacterial cells in which they grow. They may carry genes that confer advantages on their host cell, such as resistance to an antibiotic. The host cell in turn allows the plasmid DNA to be replicated to a limited extent in the cell, thereby ensuring the continued presence of the plasmid in the daughter bacteria arising when a parent bacterium divides.

Some phage and plasmid DNA's are (like SV40 DNA) small in size, ranging in complexity from several thousand to 50,000 bases. Because of their small size they can be manipulated and restructured by a variety of recently developed tools. The molecules are easily isolated, unbroken and in large amounts. They can be cut at a number of defined sites with restriction enzymes and the resulting fragments can be rejoined with one another or joined to foreign DNA segments to reconstitute the original molecule or make a hybrid molecule. The rejoining is done with readily available enzymes of bacterial origin known as DNA ligases, which recognize the ends of DNA molecules and fuse them without leaving any trace of the joining.

A hybrid DNA made of a plasmid fused with foreign (say mammalian) genetic material can replicate when it is introduced into a bacterial cell. This means the plasmid genome can serve as a "vector" for establishing and amplifying the foreign DNA in bacteria. A phage vector functions similarly, and it can serve as well to convey the foreign DNA from one bacterium to another. When the vector DNA is copied in the course of replication, the inserted foreign DNA is copied too.

The process of cloning begins with whole cellular DNA of an organism such as a mammal. The DNA is cleaved into fragments of a size (from about 1,000 to about 30,000 bases) that can be accommodated by the carrying capacity of one or another vector. A complex genome such as the human one can be broken down into a few hundred thousand DNA fragments. Each fragment can be separately inserted into a vector DNA molecule. The process does not require painstaking molecule-by-molecule assembly by a patient technician. Instead millions of insert and vector DNA molecules are mixed together and the process is completed in minutes by the addition of DNA ligase. If the resulting collection of hybrid molecules is large enough, any single gene of interest will surely be found embedded in one or another of the DNA segments linked to the vector molecules.

Each of these hybrid molecules, part vector and part inserted mammalian DNA, can now be introduced into a bacterial cell, where they are replicated many times over; each hybrid molecule spawns a separate progeny population, all of whose members are identical with the founder. Such a population is often called a clone to reflect its descent from a common ancestor and the identity of all its members.

The term "clone" has acquired another meaning. It is applied specifically to the bits of inserted foreign DNA in the hybrid molecules of the population. Each inserted segment originally resided in the DNA of a complex genome amid millions of other DNA segments of comparable size and complexity. When the manipulations described above are completed, the same segment is present in pure form within the confines of the particular clonal population, contaminated only by the associated vector DNA. The inserted DNA segment has been isolated from its previous surroundings and selectively amplified: it has been cloned, and so the purified DNA insert itself is often called a "clone."

The process of cloning requires one further step, which is usually the most challenging of all. The insertion and amplification process has given rise to hundreds of thousands of different clonal populations, each descended from a single hybrid DNA molecule. If the initial hybrids were properly diluted before being amplified, each descendant clonal population is physically separated from other populations carrying different inserted DNA's. Having established this large array (a "library") of distinct clonal populations, the experimenter is now faced with having to identify the one or several populations carrying the inserted DNA of interest.

Identification can be simple if a related gene or DNA segment has been cloned before. The previously cloned DNA can be labeled with a radioactive isotope; under appropriate conditions the radioactive DNA will preferentially stick to the clone of interest (because complementary DNA strands "hybridize" by base pairing) and thus identify it. The most interesting cloning is done, however, to isolate genes that have never been cloned before, even in related form. A variety of clever strategies have been developed to address this challenge. The goal is to develop a specific probe with which to scan a library of clones and identify the clones of interest.

One strategy for probe development depends on the fact that some proteins are expressed at a high level in specialized cells. In the precursors of red blood cells, for example, globin (the

protein component of hemoglobin) is synthesized in much larger amounts than any other protein. The mRNA that directs its synthesis is also present in large quantity, and there are ways to isolate it readily from other mRNA's in the same cell. The isolated mRNA, or a DNA copy of it, can serve as a probe that will hybridize with the corresponding gene sequence in a genomic library. Sophisticated versions of this strategy allow the mRNA encoding a protein of interest to be isolated selectively from a thousandfold excess of other mRNA molecules present in the same cell.

Often the protein whose gene is sought is rare, so that its mRNA cannot easily be isolated. In such cases a small amount of the protein is purified and the amino acid sequence of some part of it is determined. Knowing the genetic code, one can back-translate the amino acid sequence to learn what DNA base sequences are likely to be present in the gene encoding the pro-

tein. Small pieces of DNA corresponding to these derived base sequences can be synthesized by assembling off-the-shelf nucleotides. These man-made gene fragments serve as probes for identifying the clone of interest.

Yet another strategy that begins with a protein depends on antibodies directed against the protein whose gene one seeks to clone. Bacteria infected by a phage carrying the wanted gene synthesize small quantities of the protein, and so a phage library can be screened by the proper antibody, which binds to the protein and thereby identifies the gene-carrying clone.

With these and other experimental strategies available, the technology is now at hand to clone any gene whose protein product is known and can be isolated in even a small amount. Given sufficient interest, any of the genes encoding the many hundreds of enzymes that have been studied by biochemists can be isolated. The genes for the important structural proteins of the cell,

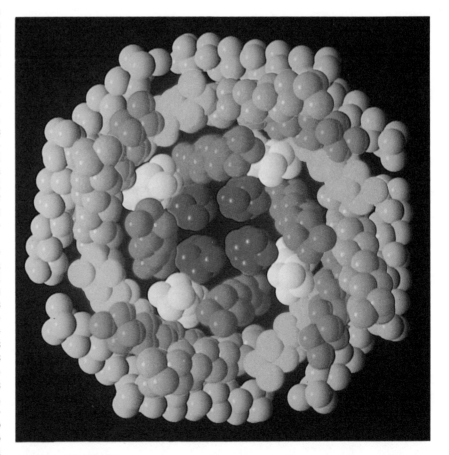

SODIUM CHANNEL, a large protein molecule embedded in the membrane of nerve cells, has been modeled by H. Robert Guy of the National Cancer Institute; this image of one model was computer-generated by Richard J. Feldmann of the National Institutes of Health. The protein admits sodium ions to the neuron, thereby supporting the action potential, the voltage pulse that ultimately triggers the release of neurotransmitter. The protein has four homologous domains; each domain includes eight distinct protein substructures. Similarly colored groups of spheres represent the four homologous versions of each substructure. (Two substructures in each domain are very similar and are both shown in pale green.)

including those that determine cellular architecture, have been cloned. Other genes encoding such intercellular messengers as insulin, interferon, the interleukins and a number of growth factors have been isolated. Indeed, genes are being cloned and their sequences deciphered faster than the new data can be fully interpreted. Most of the sequences are now being stored in computer banks; perhaps future generations of biologists, aided by new analytical procedures, will be able to interpret them fully.

The genes specifying known proteins account for only a small part of a complex organism's total genetic repertoire. Most of the remaining genes probably encode proteins too, but so far their existence is implied only by the effects they exert on cellular and organismic structure and function. Some of them specify biochemical conversions in the cell, others govern complex developmental processes that create shape and form in a developing embryo and still others may specify behavioral attributes of an organism. Such genes remain elusive because the means of identifying them in genomic libraries are limited.

The flow of genes from genome to gene library makes more things possible than the detailed description of DNA and protein structure. Once cloned, a gene can be inserted into a foreign cell, which can be forced to express it. The cell then synthesizes the protein the gene specified in its original home.

The gene to be expressed is excised from the vector in which it was cloned and subjected to important modifications. The modifications are necessary because a mammalian gene carries regulatory sequences that promote its transcription into mRNA in its home cell, not in a bacterial cell. These need to be replaced by bacterial regulatory sequences. The modified gene is then introduced into an "expression vector": a plasmid designed to facilitate the expression of the gene in a foreign cellular environment. The mammalian gene (or a similarly engineered plant gene) carrying bacterial regulatory sequences is then introduced into a selected foreign host, usually a bacterial or yeast cell.

A protein that is synthesized only in limited amounts in its normal host can be produced in large quantity when its gene is redesigned for high-level expression in bacteria or yeast. This can confer great economic advantage and represents a cornerstone of the biotechnology industry. Microorganisms bearing cloned genes can be grown quite cheaply in large volume in fermentation chambers, leading to an enormous scale-up in protein production. Among the products currently being manufactured or being considered for manufacture are insulin, interferon (for combating infections and perhaps tumor growth), urokinase and plasminogen activator (for dissolving blood clots), rennin (for making cheese from milk), tumor-necrosis factor (for possible cancer therapy), the enzyme cellulase (to make sugar from plant cellulose) and viral peptide antigens (for creating novel and safe vaccines).

In a different version of gene insertion, mammalian genes that have been cloned in bacteria are introduced into mammalian cells grown in culture rather than into microorganisms. Although cultured mammalian cells cannot be grown economically in the large numbers that characterize a bacterial or yeast culture, they have the advantage of being able to make minor but significant modifications to proteins encoded by mammalian genes. For example, certain mammalian proteins function better when sugar and lipid side chains have been attached to their amino acid backbone. The addition takes place routinely in mammalian cells but not in bacteria.

Cloned genes can now be inserted not only into microorganisms or cultured mammalian cells but also into the genome of an intact multicellular plant or animal. Here the motives are quite different from those governing the genetic engineering of unicellular microorganisms to achieve large-scale production of desirable gene products. Plants and animals can be modified genetically in an effort to alter such organismic traits as growth rate, disease resistance and ability to adapt to novel environments.

Gene insertion into a multicellular organism is a quite different project from gene transfer into a single cultured cell. The introduction of a cloned gene into most types of cells in a plant or an animal can alter the behavior of only those few cells that acquire the gene. Obviously it is of far greater interest to imprint the change on an entire organism and on the organism's descendants. That calls for gene insertion specifically into germ cells (sperm or eggs), which transmit genetic information from parent to offspring.

Techniques are indeed now available for achieving germ-line insertion into mammals, flies and certain plants. It is done either by direct physical injection of a cloned gene into the early embryo or by the use of a viral vector to carry the gene into the cells of an embryo. Again the resulting animal (or plant) carries the inserted gene in only some of its cells, but now one can hope the gene is in some of its germ cells. The presence of the gene in germ cells may allow some of the organism's offspring to inherit the inserted gene along with other parental genes, so

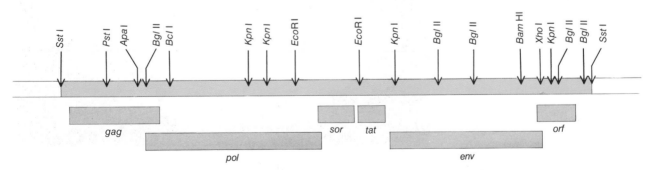

**RESTRICTION-ENZYME MAP** of the genome of the virus HTLV-III, the AIDS agent, was developed in the laboratory of Robert C. Gallo of the National Cancer Institute. Such a map is the primary means by which molecular biologists depict the organization of a stretch of DNA. The DNA is cleaved with a restriction endonuclease, an enzyme that cuts DNA at specific sites. The size of the fragments is known from the distance they travel through an electrophoresis gel. Individual fragments can be cloned and sequenced. Cleaving a genome with a number of different restriction enzymes provides additional mileposts. This relatively simple map of HTLV-III shows the sites where several restriction enzymes cleave the HTLV-III DNA (*top*) and the locations of the various genes (*bottom*). For example, the surface antigen of the virus is encoded by the *env* (for envelope) gene, and the enzyme (reverse transcriptase) that copies the RNA of the virus into DNA is encoded by *pol* (polymerase). The total length of this DNA is 9.3 kilobases (thousands of bases).

that it will be present in all their cells. Thus incorporated into the germ line of the progeny organisms, the gene is passed on to, and affects, the descendants of those organisms.

Techniques for germ-line insertion are still limited in important ways, and they may forever be. They cannot direct the foreign DNA to insert itself (to "integrate") into a particular chromosomal site; the locus of insertion is random. They cannot supplant an existing gene in the organism by knocking it out; rather, they merely add incrementally to the existing genome. Moreover, inserted genes do not always function precisely like resident genes, which are turned on and off at appropriate times in development.

Germ-line insertion is nonetheless powerful. Mice have been developed that carry and transmit to their offspring the genes for extra growth hormones. Giant mice (half again as large as normal) ensue; cattle with altered growth properties will soon follow. Flies have been developed that carry a variety of inserted genes, leading to novel insights into the molecular biology of fly development. Plants are being developed that carry genes conferring resistance to herbicides. As gene-insertion techniques are improved and as additional genes are cloned, the possibilities for altering organismic traits will expand enormously. The molecular biologist will no longer confront living forms as the finished products of evolution but will be an active participant in initiating organismic change.

For experimental biologists, cloning and its associated techniques have attained and will retain a preeminent role. Cloning makes it possible to analyze a biological system at three levels. First, the genes relevant to a particular biological problem can be isolated, the sequence of the DNA can be elucidated and the functioning and regulation of the DNA can be revealed. Second, once the DNA of a gene has been cloned the RNA transcribed from it can be produced in large amounts for study. The RNA can act in many ways to modulate the expression of genes; RNA structure and processing are central to a full understanding of gene function. Third, what is perhaps the greatest advantage of cloning stems from the analysis of the proteins encoded by a gene. How do various proteins act to elicit myriad responses in the cell? Proteins that formerly were available for study in minute amounts can now be made in great quantities once their gene has been isolated. In sum, all the major macromolecular components of a bio-

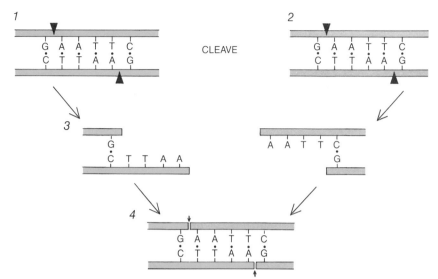

CLONING IS FACILITATED by "sticky ends" generated by some restriction enzymes. The enzyme *Eco*RI, for example, makes a staggered cut in the sequence *GAATTC*. When genomic DNA (*1*) and a vector DNA (*2*) are cut with *Eco*RI, the ends of the resulting fragments have single-strand projections of complementary bases (*3*). When the fragments are mixed, hydrogen bonds (*dots*) form between those bases, reversibly joining genomic and vector DNA's (*4*). The joint is sealed irreversibly with the enzyme DNA ligase (*arrows*).

logical system can now be made available in large amounts, in pure form.

Equally important is a newly gained ability to perturb biological systems. Genes and their encoded proteins can be redesigned so that new functions can be imparted to DNA and proteins. The relations among the interacting components of a biological system can be altered to generate novel and often revealing behavior by the system as a whole. The redesigning of genes is accomplished by changing DNA sequences through what is termed site-directed mutagenesis. This may involve the replacement of one restriction-enzyme fragment with another in the midst of a cloned gene. Alternatively, chemically synthesized DNA segments may be stitched into a gene, replacing or adding to existing sequence information. Single nucleotides can be substituted as well to create point mutations, the most subtle changes a gene can undergo. Genetic changes that have accumulated in a gene over hundreds of millions of years of natural evolution can be mimicked and superseded by several weeks' manipulation in the laboratory.

Genes altered by these techniques can then be reintroduced into the biological systems with which they normally interact. An enzyme having a low affinity for the substrate on which it acts can be engineered to associate avidly with the substrate or even to redirect its attention to novel compounds. A protein that normally is transported to one cellular compartment can be redirected to other sites in

the cell. A gene that normally is stimulated to expression by one agent can be made to respond to a completely new signal. In short, by altering the genes that organize a biological system the molecular biologist can change the usual relations between its elements in ways that show how the system normally works. Many biologists of the future will think of a biological system in terms of a series of well-defined mechanical parts that can be dismantled, engineered and reassembled under the guidance of the molecular mechanic.

It is still far from clear that attempts to reduce complex systems to small and simple components, pushed to an extreme, can provide adequate insights for coming to grips with the great problem biologists confront today: describing the overall functioning of a complex organism. Can the biology of a mammal be understood as simply the sum of a large number of systems, each controlled by a different, well-defined gene? Probably not. A more realistic assessment would be that the interactions of complex networks of genes, gene products and specialized cells underlie many aspects of organismic function. Each gene in an organism has evolved not in isolation but in the context of other genes with which it has interacted continuously over a long period of evolutionary development. Most molecular biologists would concede that they do not yet possess the conceptual tools for understanding entire complex biological systems or processes having multiple interacting components—such as, for

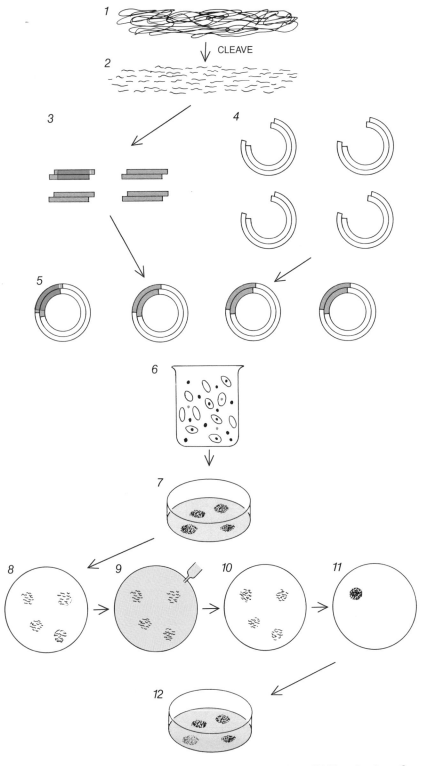

example, the process of embryonic development.

Gene cloning has already illuminated another corner of biology: the history of evolutionary development. Human beings, who appeared in their present form only several hundred thousand years ago, are rapidly becoming privy to some of the evolutionary events that, as long as one and two billion years ago, began to shape the life forms that now populate the earth. The steps that generated the first cellular life forms may never be known, but many of the subsequent changes, memorialized in the DNA of present-day organisms, are being identified by means of the cloning and sequencing of genes. The relatedness of organisms can be established with little ambiguity simply by analysis of their cloned DNA segments. Sophisticated computer programs have been mustered to help analyze these sequences and determine evolutionary relations.

What makes it possible to reconstruct the genealogy of life is the remarkable conservation of certain ancestral sequences over very large evolutionary distances. This conservation has been of great practical advantage to the molecular biologist for another reason. Genes and associated biological problems that are difficult to attack in one organism can be resolved in another organism that is more amenable to manipulation.

There are, for example, certain oncogenes (cancer genes) whose functions are analyzed only with difficulty in human cancer cells. Closely related genes have been identified in the DNA of yeast—an important finding in itself because it indicated that the genes played essential roles in normal cellular physiology long before the appearance of multicellular organisms. The relative simplicity of the yeast cell and the elegant manipulations to which it can be subjected have made experiments possible that are beyond the reach of those working with mammalian cells. The experiments first yielded information on how the oncogenes function; extrapolation of the results to mammalian cells will contribute to the rapidly evolving understanding of the molecular basis of cancer.

The application of cloning techniques to the study of evolution affects understanding of the human species as well. The human species, like others, is composed of a genetically heterogeneous collection of individual organisms. This diversity provides the seeds of future evolution, and the particular versions of genes carried by some people will confer advantage on human-

**CLONING OF A GENE is the central operation of recombinant-DNA technology. One approach is diagrammed. The DNA of a mammalian genome (1) is cleaved with a restriction enzyme. Some of the resulting fragments (2, 3) may include the gene of interest (color). Many copies of a plasmid cloning vector are cleaved with the same enzyme (4). Plasmids and genome fragments are mixed and joined by DNA ligase (5), and recombinant plasmids are introduced into bacteria (6). The bacteria are plated in a culture dish thinly enough so that each resulting colony (7) is a pure clone whose members are descended from a single cell. The cells of a few clones may contain a recombinant plasmid carrying the gene of interest. There is an easy way to find such a clone if this wanted gene's messenger RNA has been identified and can serve as a probe. A sample of the colonies is transferred to a disk of filter paper; the cells are broken open to expose their DNA (8). The RNA probe, labeled with a radioactive isotope, is added (9). It anneals only to the wanted DNA, and unannealed probe is removed (10). When the filter paper is covered with a photographic emulsion, the radioactive probe makes a spot on the emulsion (11), identifying the clone of interest (12).**

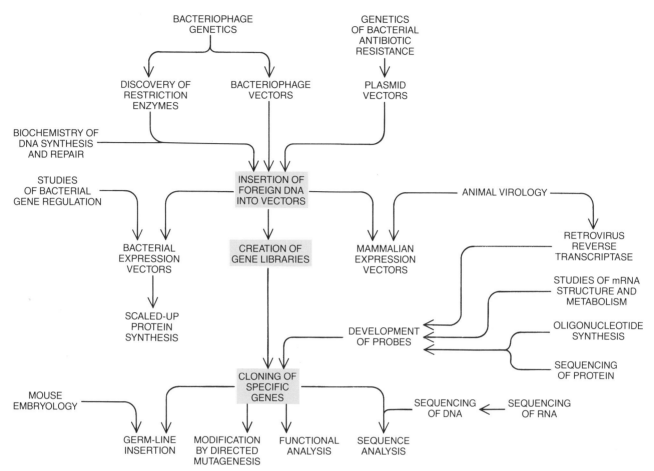

**LINE OF DESCENT** leading to contemporary recombinant-DNA technology is traced in this flow diagram. It shows how a wide variety of studies and findings contributed to a succession of new techniques, which led in turn to new capabilities that opened up new areas of study. The evolution of technology has been facilitated by the continual development of novel tools and methods.

kind in future encounters with the forces of natural selection. Versions of certain other genes present in the human gene pool are clearly disadvantageous at present: genes that predispose toward sickle-cell disease, atherosclerosis, cancer, hemophilia and a host of other metabolic disorders. The genes implicated in these diseases are rapidly being identified and cloned, and with this cloning comes the prospect of clear diagnosis of genetic predisposition in adults and even in the embryo.

The recognition of genetic diversity and genetic lesions within the human gene pool has stimulated research on ways of repairing defective genes, both in afflicted individuals and in their descendants. Techniques are now available for introducing cloned genes into certain somatic (non-germ-line) tissues such as bone marrow and skin cells. The cloned genes can be intact, healthy versions of genes present only in defective form in the cells of affected individuals. Such gene transfer into somatic cells may reverse, at least partially, the effects of certain genetic lesions.

Insertion of cloned genes into the human germ line as well can be contemplated within this decade; the hope is thereby to cure a genetic defect in the descendants of an afflicted individual.

The prospect of such therapeutic intervention has triggered heated debate. There are two broad issues. The first one concerns the human germ line itself. In tampering with it, do human beings cross over an inviolable boundary? Should the human germ line be ringed by a wall to protect its sanctity? And will initial attempts to ameliorate obviously disadvantageous conditions soon be replaced by more ambitious plans to "improve" the human germ line? Genes affecting aspects of intelligence, disposition and body build will surely be identified in the next decade or so, and they could become tempting targets for manipulation.

The other issue transcends the human condition. Is germ-line alteration in general a threat? As described above, genomic alteration is now practical in bacteria, flies, plants and mam-

mals. How will existing ecological interrelations be perturbed by the presence of organisms carrying altered genes in their germ line? The results to date have been reassuring, in that genetically altered organisms have proved to be less viable than their wild-type counterparts and hence unable to affect existing ecosystems substantially. A number of arguments can be mustered, each of which persuades that ecological imbalance is unlikely—and none of which provides total assurance that accidents are impossible.

A profound disquiet underlies consideration of these issues. As physicists did 40 years ago, contemporary biologists have invaded a domain of human innocence. Is life to be redesigned to suit human needs and curiosity? Can—and should—life be described in terms of molecules? For many, such description seems to diminish the beauty of nature. For others of us, the beauty and wonder of nature are nowhere more manifest than in the submicroscopic plan of life described in part in the pages of this book.

# 2

# DNA

# DNA

*The genetic material's double helix, the fundamental molecule of life, is variable and also flexible. It interacts with regulatory proteins and other molecules to transfer its hereditary message*

by Gary Felsenfeld

The double helix of DNA is the most familiar symbol of the biological revolution that began in the 1940's. The model of the DNA molecule proposed by James Watson and Francis Crick in 1953 had a particularly strong impact because the structure contained within itself indications of how DNA might perform its function of storing and transmitting genetic information. Much of the explosive advance in molecular biology set in motion by that discovery has been aimed at understanding how DNA interacts with the other components of the living cell to express the information it encodes. Some recent findings of this line of research have led to the realization that, although DNA seems in its familiar double-helical form to be fairly inert and inflexible, it is in fact both chemically and structurally one of the most versatile of molecules. Indeed, it must be to carry out its many functions.

The exterior of the DNA double helix (often called the duplex) is dominated by the backbones of the two interwound strands. Each strand, a long polymer of subunits called nucleotides, has a backbone in which phosphate groups alternate with a sugar, deoxyribose, to form a covalently linked chain: a structure held together tightly because atoms share pairs of electrons. The chains have polarity, or direction: in the duplex they run parallel but in opposite directions, with a right-handed twist.

Attached to the sugar ring of each nucleotide is one of four bases: adenine ($A$), guanine ($G$), thymine ($T$) or cytosine ($C$). (Adenine and guanine are called purines; thymine and cytosine are pyrimidines.) The order of the nucleotide bases along the chain provides the information that specifies the composition of all the organism's protein molecules.

The bases project from the sugarphosphate backbone toward the center of the structure. Each base is linked by either two or three hydrogen bonds to a base on the opposite strand. Given the structure and size of the bases, this arrangement requires that an $A$ on one strand be paired with a $T$ on the other and that $G$ be paired with $C$. The nucleotide sequence on one strand thus determines the sequence on the other.

Looking at the Watson-Crick model, one may find it difficult to imagine the structure as a reactive molecule. Its stacking of the planar base pairs on one another makes the double helix rather stiff. The disposition of the bases—the variable elements—toward the interior suggests a structure designed to preserve the genetic code but not one that might easily take part in biochemical reactions. This appearance is illusory: the double helix is actually capable of assuming many forms and of reacting in many different ways with other molecules in the cell. These properties are exploited by the cell to control the way in which the genetic information held in DNA is expressed.

Even at the time of Watson and Crick's discovery it was evident from X-ray-diffraction patterns of DNA fibers that DNA could exist in at least two forms: $B$ (the form Watson and Crick discovered) and $A$, wherein the base pairs have a different tilt and are displaced outward with respect to the axis of the helix. In recent years it has become clear that the $A$ and $B$ forms revealed by the analysis of fibers are average, approximate structures. Detailed X-ray studies of crystalline DNA with defined base sequences reveal that within the "$B$ family," for example, there is a large sequence-dependent variation in local conformation. Other studies of DNA fragments in solution suggest that some kinds of nucleotide sequence are capable of imparting a permanent bend to the duplex. Every DNA sequence is thus recognizable from the outside of the duplex not only through the identity of its bases but also through variations in the details of positioning of both the bases and the backbone.

More unusual DNA sequences have also come to light. Certain sequences of alternating purines and pyrimidines are able to undergo conversion from a normal right-handed $B$ helix into the left-handed $Z$ form, which was first identified by Alexander Rich of the Massachusetts Institute of Technology and his colleagues. The backbone of $Z$ DNA follows a more irregular path than the backbone of the $B$ form, and the two grooves (major and minor) along the flanks of the $B$ helix are replaced by a single deep minor groove; the atoms defining the major groove of $B$ DNA are moved to the surface of the $Z$ duplex.

Other special DNA sequences are also capable of assuming distinctive conformations. One is the inverted repeat, in which a sequence of bases is

**REGULATORY MOLECULE operating on the DNA of the bacterial virus called lambda is the *cro* repressor. It is the protein structure shown at the right in the image on the opposite page, slightly separated from the schematic representation of the DNA double helix at the left. In the repressor each amino acid is represented by one sphere: red for positive charge, blue for negative charge, white for hydrophobic and green for hydrophilic. In the DNA the orange spheres represent components of the sugar-phosphate backbone of the helix; the yellow spheres follow the bottom of the helix's "major groove." The repressor, which by a complex reaction prevents the expression of a gene, is about to enter the major groove, where its amino acid side chains will make hydrogen bonds with the appropriate DNA bases. This bonding accounts for the specificity of the reaction. Brian W. Matthews and Douglas H. Ohlendorf of the University of Oregon made the computer-generated image.**

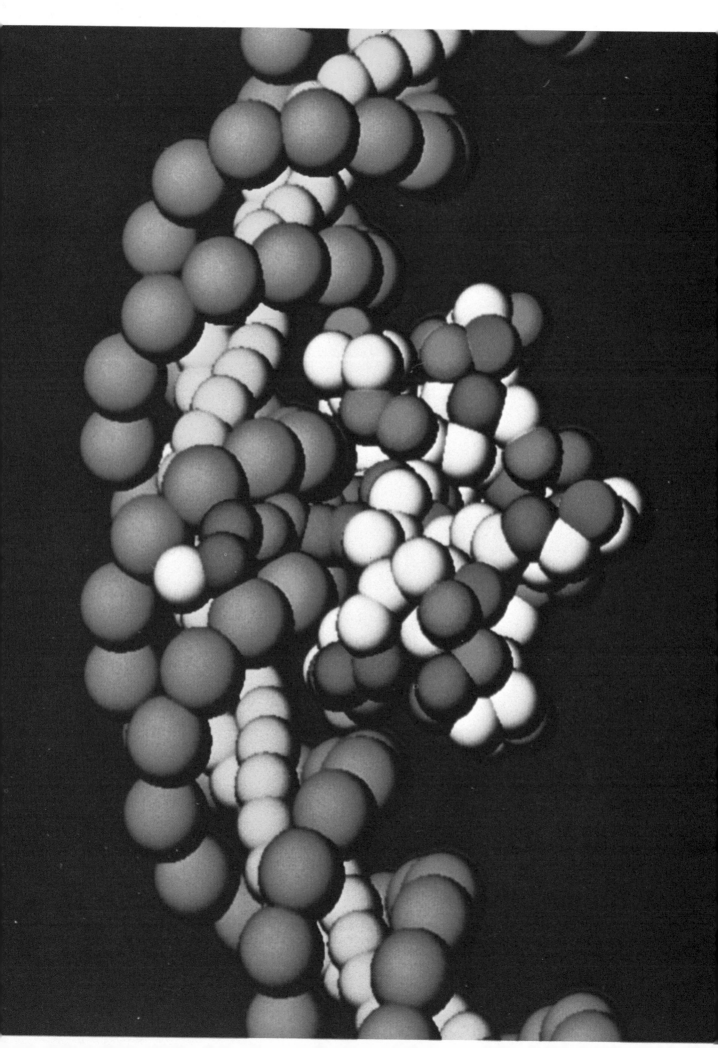

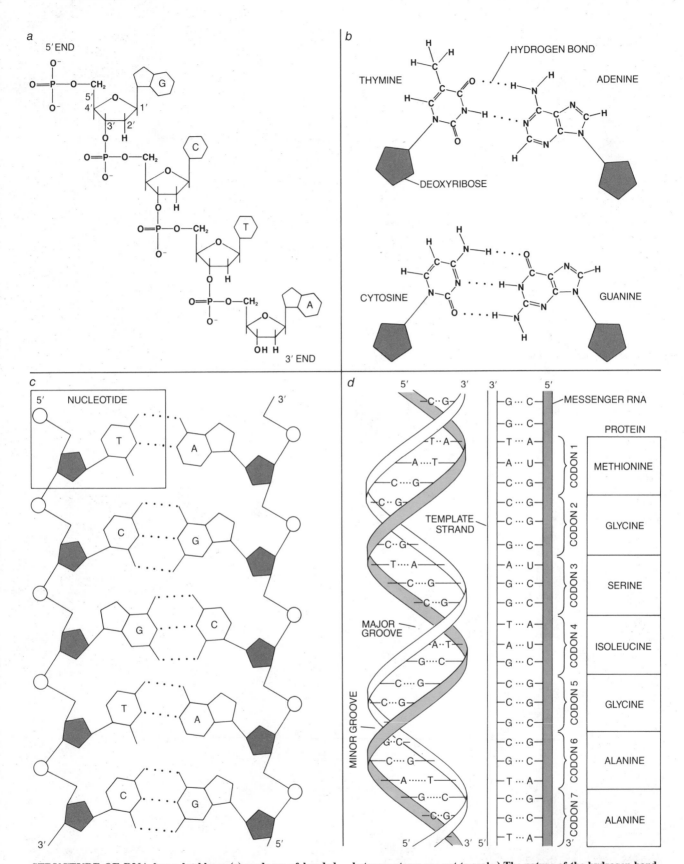

**STRUCTURE OF DNA** has a backbone (*a*) made up of bonded sugar and phosphate groups, to each of which is attached one of four bases: guanine (*G*), cytosine (*C*), thymine (*T*) or adenine (*A*). The phosphate group is represented by the structures with the *P* at the center, the sugar by the pentagon with an oxygen atom (*O*) at the top. A phosphate group connects the 5′ carbon atom of one sugar to the 3′ carbon atom of the next. The combination of sugar-phosphate group and base constitutes a nucleotide. (The distances between atoms are not to scale.) The nature of the hydrogen bonding of the bases is such that thymine always pairs with adenine and cytosine always pairs with guanine (*b*). The resulting structure is shown in two dimensions (*c*) and in three: the double helix (*d*). In conveying the genetic message of DNA the sequence of the coding strand (*color*) is transcribed into a strand of messenger RNA, which serves to make a variety of proteins. The *U* in the strand of messenger RNA stands for uracil, the RNA counterpart of thymine.

followed on the same strand by its complementary sequence in reverse order. If the strands of the double helix are separated from each other, a locally base-paired structure resembling a hairpin can form within each of the separated strands. (Because the strands are complementary, an inverted repeat on one strand implies an inverted repeat on the other.) The entire structure is thus cruciform in appearance. Since the formation of both Z DNA and cruciforms involves the unwinding of B DNA, any stress on DNA that tends to favor unwinding will also favor these unusual structures. It has been suggested that such structures may serve as recognition signals that are important for the biological role of DNA.

DNA functions as the source of genetic information by interacting with proteins that copy it into a strand of a similar nucleic acid, RNA, and with other proteins that modulate the copying activity. In the process called transcription a class of complex enzymes, the RNA polymerases, construct an RNA copy of the DNA sequence of one strand (the coding strand). Some of the RNA plays a structural role of its own, but much of it is converted into messenger RNA (mRNA), which directs the synthesis of all the cell's proteins. In every kind of cell decisions must be made as to which genes will be expressed—transcribed into RNA for subsequent translation into protein—and when. Much of this control is exerted at the level of RNA-polymerase binding.

The polymerase initiates its binding at special nucleotide sequences that are just outside the beginning of each gene or gene cluster to be copied; the region is called the promoter. (The polymerase "reads" the gene by starting at the end designated 5' and proceeding toward the 3' end.) Two questions arise at once. How might the binding strength of the polymerase be regulated? How does a protein such as polymerase distinguish specific gene sequences?

A few general mechanisms for polymerase binding come to mind. It could be inhibited if the binding site were blocked by another protein or if the local structure of the DNA were altered. On the other hand, polymerase binding might be enhanced by different kinds of changes in the DNA structure or by the binding of regulatory proteins close enough to interact favorably with the polymerase and stabilize its attachment to its initiation site.

Organisms resort to all these strategies in controlling the expression of their genes. The mechanisms of regulation require many kinds of special nucleotide sequences and many proteins to recognize them.

The most detailed data concerning regulatory processes come from the study of bacteria and viruses. In one recurring pattern that emerges from such studies the action of RNA polymerase is modified by the binding of specialized regulatory proteins to DNA near the point of initial attachment of the polymerase.

The classic example of such a mechanism is the *lac* operon of the bacterium *Escherichia coli*. The operon is a cluster of genes associated with the metabolism of the sugar lactose; the first of the genes codes for the enzyme beta-galactosidase, which splits lactose into two smaller sugars. The rate of synthesis of the enzyme is directly coupled to the amount of lactose in the cell: the greater the amount of lactose, the greater the amount of enzyme production.

The regulatory agent responsible for this behavior is a protein molecule, the *lac* repressor, that is able to bind tightly to a sequence of nucleotides (termed the operator) situated between the binding site of RNA polymerase (the promoter) and the region coding for beta-galactosidase. In the absence of lactose the repressor is bound to the operator and blocks the attachment of polymerase. The repressor is also able to form a complex with a derivative of lactose that serves as an "inducer" because the complex binds to the operator site much more weakly. In the presence of excess lactose the complex is formed, the repressor is released from the DNA, the polymerase binds more readily to the promoter and the transcription of the gene is by this means induced.

The binding of the repressor protein to its specific site on DNA is extremely tight. It is also selective: under physiological conditions the equilibrium constant (a measure of binding strength) is about a billion times greater for binding to the operator than it is for attaching to a typical DNA sequence elsewhere in the genome. Even so, the equilibrium constant for random binding is large enough to guarantee that excess molecules of the repressor within the cell will be bound somewhere on the DNA.

This nonspecific binding also plays an important part in the regulatory process. As Peter H. von Hippel of the University of Oregon and his colleagues have noted, it serves as a competitor, keeping the repressor's specific binding—to the *lac* operator—within bounds, so that it can be reversed when the situation calls for the production of beta-galactosidase. The formation of the complex with nonspecific DNA also greatly increases the rate at which the specific repressor-operator complex can be assembled, since the repressor, once it is bound anywhere on the single DNA molecule of the bacterial chromosome, has a much increased probability of encountering the operator region. Von Hippel has demonstrated that the *lac* repressor finds the operator largely by sliding along the DNA.

The repressor protein is held to the DNA mainly by electrostatic forces and by hydrogen bonding. Electrostatic forces involve the interaction of positively charged amino acids in the protein with the negative charges in the backbone of the DNA. Although these forces contribute greatly to the strength of binding, they are by nature largely nonspecific. The specificity that allows the repressor to recognize preferentially the particular nucleotide sequence defining the operator arises largely from the formation of specific hydrogen bonds. They link a defined set of amino acid side chains in the protein with a corresponding set of hydrogen-bond donors and acceptors on the operator's nucleotide bases.

Although work on the structure of the *lac* repressor is not complete, a great deal of information has been obtained from X-ray-diffraction studies of related repressor proteins. One such structure, determined in the laboratory of Brian W. Matthews of the University of Oregon, is the *cro* repressor of the bacterial virus lambda. It takes part in a regulatory process somewhat like that of the *lac* repressor but more complex. The *cro* protein has alpha-helical segments arranged in such a way that they are able to fit into the major groove along the DNA duplex and make the appropriate hydrogen-bond contacts [*see illustration on page 15*]. As Matthews has pointed out, just such an arrangement of alpha helixes is common to a number of regulatory proteins of bacteria and viruses. The specificity exhibited by each protein for a particular sequence of bases derives from the array of particular amino acids in the hydrogen-bonding positions. Although other new sequence-specific DNA-binding proteins are quite different in structure, the general principles of recognition are likely to be the same.

How does the variability of DNA's conformation enter into interactions of this kind? DNA in its various forms displays large local excursions in structure that depend on the details of the nucleotide sequence. Probably proteins that recognize a particular hydrogen-bonding pattern within a groove of the double helix take account of these local peculiarities of structure to some extent in the placement of their reactive sites. Given the structural plasticity of DNA, it is also likely that the DNA deforms somewhat to accommodate the protein. The final structure is a compromise that depends on the size of the forces necessary to deform the duplex.

In the case of the *lac* and *cro* repressors the answers to these questions await X-ray-diffraction studies of the complex between the protein and its DNA target. Studies involving other proteins, however, have already demonstrated that in certain cases the double helix can be considerably deformed. One such distorted complex is formed between the *lac* operon and another regulatory protein called CAP (for catabolite activator protein).

CAP is a protein that stimulates the transcription of a number of genes in response to the presence of increased concentrations of a small effector molecule, cyclic adenosine monophosphate. When cyclic AMP binds to CAP, the complex is able to bind to DNA. In the case of the *lac* operon the point of attachment is just upstream (toward the 5′ end) from the binding site of RNA polymerase. The shape of this complex in solution has been studied by Donald M. Crothers of Yale

University. He finds that although the *lac* DNA is straight when it is protein-free, the attachment of CAP induces a bending that results in anomalous migration in gel electrophoresis, the technique of "sieving" whereby molecules migrate through a porous matrix under the influence of an electric field and are sorted according to size and shape. Crothers suggests the bending may serve to bring CAP in direct contact with the adjacent polymerase, thus conveying the stimulatory signal.

Another important way in which the DNA structure is involved in the regulation of transcription is through the phenomenon termed supercoiling. Imagine a laboratory procedure in which a linear fragment of DNA is bent, while lying flat on a surface, until opposite ends touch, so that the strands can bond covalently to form a closed circle. Before sealing them together the experimenter untwists the two strands of DNA by a few turns so that after sealing there is a driving force that tries to restore the normal twist. Because of the topological constraints, however, any attempt to restore the normal twist must be accompanied by a distortion that will keep the DNA backbone from lying in the plane. The DNA is now supercoiled, that is, the double helix itself follows a superhe-

lical path. Structures formed by untwisting the DNA duplex before joining are called negatively supercoiled.

Supercoiled DNA is not merely a laboratory curiosity; it is the natural form of most DNA in living cells. As Abraham Worcel, now at the University of Rochester, and David E. Pettijohn of the University of Colorado Medical Center in Denver showed some years ago, the chromosomal DNA of bacteria is organized in loops held together at their base by some combination of protein and RNA, which serves to isolate each loop topologically from its neighbors, so that it behaves like a closed circle that can be supercoiled independently. It is evident that supercoiling must be important to such cells, because they expend considerable energy to generate negative supercoils and to maintain the amount of supercoiling at a rather constant level.

The enzyme that is primarily responsible for carrying out this function in bacteria is DNA gyrase, which was discovered by Martin F. Gellert of the National Institute of Arthritis, Diabetes, and Digestive Diseases and his colleagues in 1976. Gyrase is an enzyme that is able to convert a relaxed, closed, circular molecule (or a topologically isolated loop) into a negatively supercoiled form. It obtains the nec-

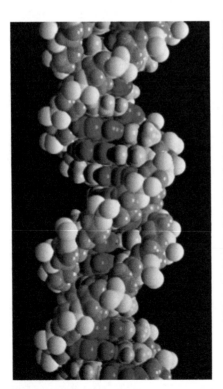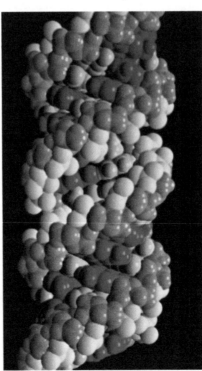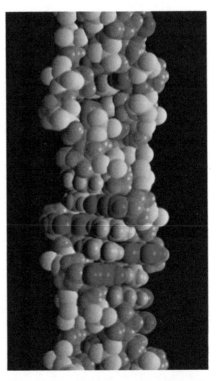

**STRUCTURAL FORMS OF DNA include** *B, A* **and** *Z.* **In these images, which Richard J. Feldmann of the National Institutes of Health generated by computer, the phosphate-sugar backbone appears in white (phosphate) and blue gray (the sugar deoxyribose). Within the structure the pyrimidines thymine and cytosine (*yellow*) on one strand are paired respectively with the purines adenine or** guanine (*red*) on the other. *B* DNA (*left*) is a right-handed helix with about 10 base pairs per turn. *A* DNA (*center*) is also right-handed, but the bases are shifted away from the axis of the helix and are inclined with respect to that axis. The recently discovered *Z* DNA is a left-handed helix with a zigzag backbone that gives the structure its name. *Z* DNA has 12 base pairs per helical turn.

essary energy for this reaction by hydrolyzing adenosine triphosphate (ATP), the major energy source for biological reactions.

Gellert and his collaborators have explored the way the level of supercoiling is modulated in *E. coli* by gyrase and other enzymes. The most remarkable finding is that the rate of transcription of the gyrase gene is itself controlled by the supercoiling level, in such a way that gyrase production is shut off as the average number of negative superhelical turns increases. Further study reveals that the transcription of many bacterial proteins is affected by DNA superhelicity (although some are made at a faster rather than a slower rate when supercoiling increases).

How could supercoiling be involved in the regulation of transcription? It is known from many studies of the action of RNA polymerase that the initiation of transcription takes place in two steps. First the polymerase binds to the promoter; then the duplex undergoes a local denaturation, or "melting," that separates the two DNA strands and thus makes it possible for transcription to proceed. Denaturation untwists DNA and thereby relieves superhelical stress. Energy considerations therefore suggest that increased negative superhelicity should favor denaturation and so make it easier for transcription to begin. Although such a simple model does not explain the regulation of gyrase synthesis, it does appear to account for the effects of supercoiling on the expression of certain other bacterial genes.

Supercoiling provides the cell with a powerful mechanism for the regulation of transcription because the overall superhelical stress of a closed circular or looped DNA can be altered by a change anywhere in the domain of such a structure. Negative supercoiling not only favors the unwinding of the DNA duplex but also stabilizes the formation of the *Z* DNA and cruciform structures I have described. These deformations and others could solve a crucial biological problem: How can regulatory elements, necessarily dispersed over considerable distances on linear DNA, communicate with the gene itself?

Nowhere is this issue more pressing than in the DNA of higher organisms. A typical plant or animal cell contains perhaps 1,000 times more DNA than *E. coli,* and the expression of that DNA must be subject to far more elaborate controls than are needed in a microorganism. One of the chief tasks of these controls is to guarantee that, in each of the specialized cells

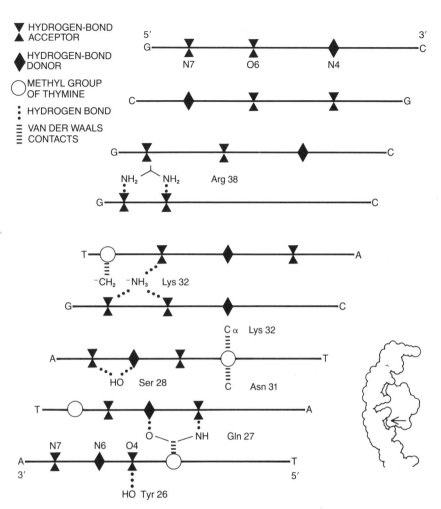

SCHEME OF BONDING between the *cro* repressor of the lambda virus and its target DNA was proposed by Matthews and his colleagues. In this view one is looking into the major groove of the DNA from the lower half of the repressor protein; the perspective is shown in the diagram (*right*) of the image on page 15. Nine base pairs of the DNA appear, beginning at the top with *GC.* Symbols (*left*) denote the nature of the bonds. Also shown are certain amino acid groups on the repressor thought to make a specific contact with an adjacent base; for example, *Arg 38* stands for an arginine that is the 38th group from the amino terminus of the repressor protein. The upper helical section of the repressor bonds similarly.

making up the differentiated tissues of the higher organism, only those genes relevant to the function of that kind of cell are turned on. To a first approximation the approach to control taken by eukaryotic (nucleated) cells is like that in microorganisms.

The thymidine kinase (*tk*) gene of the *Herpes simplex* virus provides a particularly well worked out example of the organization of regulatory sequences in eukaryotes. To identify the gene's control elements the gene is first isolated and cloned in a plasmid vector. (Plasmids are small, closed, circular molecules of DNA. Since they can multiply independently in bacteria, one can obtain large quantities of homogeneous material for study.) The expression of the gene can then be studied by introducing the plasmid into a suitable cell.

In experiments carried out by Steven McKnight of the Carnegie Institution

of Washington's Department of Embryology and Richard Axel of Columbia University the plasmids carrying *tk* are injected into oocytes, or immature eggs, of the frog *Xenopus laevis*. In the oocyte the gene is transcribed directly from the plasmid, and the amount of *tk* RNA produced can be measured.

A second kind of assay employs a powerful technique developed earlier by Axel and his collaborators. A culture of mutant mouse cells lacking their own *tk* gene is persuaded to absorb the viral DNA, which becomes integrated into the cell's DNA complement and is expressed. If the cells are then transferred to a special medium in which thymidine kinase is required for growth, the level of expression of the enzyme will be shown by the number of cell colonies that grow.

To dissect the *tk* gene's control region, which lies in the 5' direction from the sequence coding for the gene, it is necessary to couple these assays with recombinant-DNA technology. Starting at one end or the other of the region, bits of the naturally occurring sequence are systematically replaced with an equivalent length of foreign DNA. When the pattern of expression

of these deliberately constructed mutant genes is examined, several areas with distinct functions can be seen.

The first of them, about 25 nucleotides upstream from the start of transcription, appears to have an effect on the site where the synthesis of RNA

originates: when a segment containing the sequence *TATTAA* is altered, transcription levels fall and new mRNA species appear whose starting sites are displaced from the normal one. (David S. Hogness and Michael Goldberg of Stanford University first pointed out the prevalence of the sequence *TATA* and close relatives of it in the promoter region of eukaryotic genes. Its effect on the site at which transcription starts has been observed in many laboratories for many genes.)

As the defect in the sequence is moved farther in the 5′ direction, another kind of control region is encountered. It affects the total amount of transcription but not the site of origination. When this sequence is perturbed, the production of mRNA drops roughly tenfold. Another fairly homologous sequence with similar properties appears still farther upstream. Of the two sites, the one nearest the place where transcription begins seems to be critical for positive control of transcription.

Although the general approach to transcriptional regulation adopted by the herpes *tk* gene is not greatly different from that found in microorganisms, eukaryotes have also developed a variety of novel control mechanisms. One involves the stimulatory elements called enhancers, which were first identified in such animal viruses as SV40.

SV40 is a small virus whose circular DNA is long enough to code for only a few proteins. Transcription starts from each of two promoters adjacent to each other on the circle and proceeds in opposite directions. The "early" promoter, which functions at the beginning of the infectious process, incorporates some of the same elements seen in the herpes *tk* gene. It also has two copies of a 72-nucleotide sequence arranged in tandem.

Employing the kind of deletion analysis I have described for the *tk* gene, Pierre Chambon of the University of Strasbourg and George Khoury of the National Cancer Institute and their collaborators were able to show that the absence of both copies of this element reduces early gene transcription more than a hundredfold. What distinguishes this element, which has come to be called an enhancer, from the other kinds of regulatory elements I have described is that its stimulatory effect does not depend on any precise positioning with respect to the site of initiation of transcription. Walter Schaffner of the University of Zurich and his colleagues experimented by inserting the tandem repeat at varying distances from a test gene. They found it can

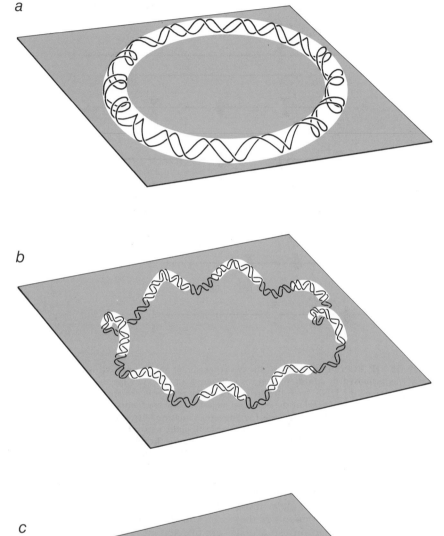

*a*

*b*

*c*

**SUPERCOILING IN DNA** changes the shape and stability of the double helix. When a helix is formed into a circle by a bond joining the ends (*a*), it lies flat in a plane because the DNA is relaxed. If the double helix is untwisted several turns before the ends are joined (*b*), it tries to resume its normal twist, and so the backbone can no longer lie in a plane. Here it has become a left-handed toroidal superhelix. A topologically equivalent form (*c*) is a right-handed supercoil that is probably closer to the shape supercoiled DNA assumes in the cell.

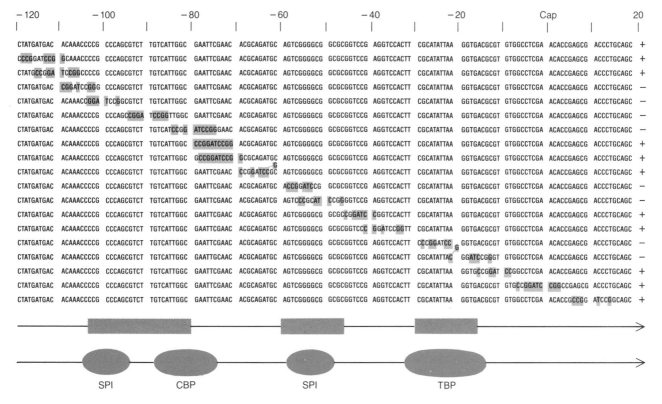

**GENE-MODIFICATION EXPERIMENTS** by Steven McKnight of the Carnegie Institution of Washington's Department of Embryology involved the thymidine kinase (*tk*) gene of the *Herpes simplex* virus. The aim was to see what effect alterations in the promoter region of the gene had on expression of the gene. The complete DNA-base sequence of the region appears in the top line. Each line below shows (*color*) the location of genetically engineered alterations in the nucleotide sequence. A plus sign at the end of a line signifies that the alteration has little or no effect on the expression of the gene; a minus sign signifies that the alteration reduces the level of *tk* output. Three regions shown by the experiment to be important to expression are indicated by the gray bars below the diagram. Recent work by McKnight in collaboration with Robert Tjian and Katherine Jones of the University of California at Berkeley has identified the binding sites of three regulatory proteins: SP1, CBP and TBP (*bottom line*). CBP and TBP respectively recognize the common regulatory sequences *CAAT* (or *CATT*) and *TATA* (which are family names for a group of similar sequences) of DNA.

exert its influence at distances of more than 3,000 nucleotides. (Enhancer sequences in the living cell are usually much closer than that to the genes they control.) As work in the laboratories of Chambon and of Paul Berg of Stanford has demonstrated, enhancer sequences exert their effect even if they are inserted with their orientation reversed.

This attribute—action independent of distance and orientation—is the defining characteristic of enhancers, which have been found subsequently in many viral genomes. Enhancers also appear in eukaryotic chromosomes. In 1980 Max Birnstiel and Rudolf Grosschedl of the University of Zurich identified an enhancer sequence near the genes specifying proteins called histones in the sea urchin; more recently such sequences have been found in association with immunoglobulin (antibody) genes and with a variety of other eukaryotic genes.

How enhancer elements work is not clear, but it is likely that their operation involves the binding of a protein. Such binding was first demonstrated for the immunoglobulin genes and

quite recently, by Chambon's laboratory, for the SV40 72-nucleotide repeat. Since action at a distance is involved, it has been suggested that the stimulatory signal may be transmitted through some modulation of DNA supercoiling. Another suggestion is that the enhancer acts as an entry site for RNA polymerase, which then travels along the DNA until it reaches a promoter. In any case, the most exciting aspect of recent work is the demonstration that enhancers are specific for cell type. It therefore seems likely that the particular protein molecule necessary to activate a given enhancer is to be found only in those cells requiring expression of the genes controlled by that enhancer.

Yet another regulatory strategy employed by some eukaryotes involves a chemical modification of the DNA by the addition of a methyl group ($CH_3$) at the carbon-5 position of cytosine. The targets of methylation in animals are almost exclusively those cytosines that are immediately followed by guanine. Not all *CG* sites are methylated, and the pattern of methyl-

ation is more or less stably inherited in any given cell type. Work in a number of laboratories has elucidated the pattern: many of the genes that are not being expressed in a cell tend to have a large fraction of their sites methylated, whereas in a cell in which the same genes are active there is a marked reduction in their level of methylation, often concentrated in the promoter region. Although the mechanism for altering methylation in this highly specific manner is not understood, it seems likely that the change is responsible in some way for helping to alter transcriptional levels.

Evidence for such a role comes from experiments showing that the promoter regions of certain genes can be inactivated by the deliberate methylation of particular sites. Aharon Razin and Howard Cedar of the Hebrew University of Jerusalem and their colleagues have demonstrated this effect by introducing DNA containing a hamster gene into mouse cells. The DNA is taken up by the cells, and under normal circumstances the gene is expressed. If the DNA is methylated beforehand, however, the gene is not expressed

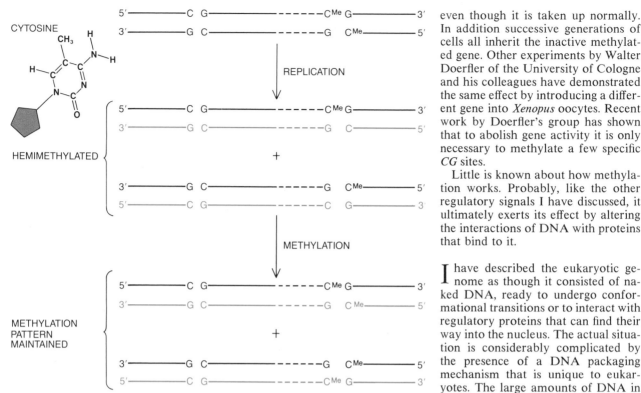

**METHYLATION OF GENES,** which entails the addition of a methyl group (CH₃) to the fifth atom of the cytosine ring, appears to be a mechanism for curtailing the expression of genes. In animals the sites of methylation are cytosine bases followed on the same strand by guanine. The pattern of methylation is passed on when a cell divides. Here the steps in replication are traced. The two strands of DNA at the top have one *CG* site fully methylated (*Me*). When new DNA is synthesized during cell division, the two strands of the DNA separate and new complementary strands (*color*) are synthesized. This process incorporates only unmethylated *C*, so that the new double strands are "hemimethylated" at the originally methylated site. In the final step a methylating enzyme acts on the DNA; it methylates *C* bases if they are opposite already methylated cytosine, thereby perpetuating the original pattern. Increased levels of methylation tend to be correlated with reduced gene expression.

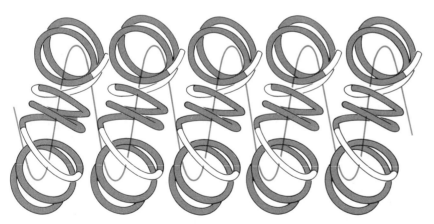

**PACKING OF DNA** in eukaryotic cells allows large amounts of DNA to fit within the cell nucleus in an orderly fashion and also provides a means whereby sequence elements widely separated in DNA may come close together for interaction with regulatory components. The compact form is called chromatin, and its fundamental unit is the nucleosome. Strings of nucleosomes are in turn folded into fibers 30 nanometers thick. A proposed model for the packing of nucleosomes in 30-nanometer fibers from red blood cells of the chicken is shown here. It focuses on the chromatosome, the central part of the nucleosome. In a chromatosome two turns of the DNA double helix (*gray*), each about 83 base pairs in length, are wrapped in a superhelix around an eight-molecule complex of the proteins called histones (not shown). In the fiber the chromatosomes, connected by linker DNA (*white*), are wound in a solenoidal array (*whose path is traced by the colored line*) with an average of six chromatosomes per turn of solenoid. The orientation of the chromatosomes is deduced from optical studies of electrically aligned fibers. The path of the linkers in this representation is speculative. For clarity only the front three nucleosomes in each turn are shown.

even though it is taken up normally. In addition successive generations of cells all inherit the inactive methylated gene. Other experiments by Walter Doerfler of the University of Cologne and his colleagues have demonstrated the same effect by introducing a different gene into *Xenopus* oocytes. Recent work by Doerfler's group has shown that to abolish gene activity it is only necessary to methylate a few specific *CG* sites.

Little is known about how methylation works. Probably, like the other regulatory signals I have discussed, it ultimately exerts its effect by altering the interactions of DNA with proteins that bind to it.

I have described the eukaryotic genome as though it consisted of naked DNA, ready to undergo conformational transitions or to interact with regulatory proteins that can find their way into the nucleus. The actual situation is considerably complicated by the presence of a DNA packaging mechanism that is unique to eukaryotes. The large amounts of DNA in the nucleus are intimately associated nearly everywhere with the proteins called histones. They fold the DNA into an ordered, compact form called chromatin.

The fundamental subunit of chromatin structure is the nucleosome. Each nucleosome contains a central part, the chromatosome, in which two turns of the DNA duplex, each about 83 base pairs long, are wrapped in a superhelix around an octamer (an eight-molecule complex) of histones. Adjacent chromatosomes are connected by a segment of relatively unconstrained linker DNA.

Nucleosomes and chromatosomes can be isolated from the nucleus of cells by partial digestion with nucleases: enzymes that preferentially cut the linker DNA and thus break the connections between the subunits. A particularly stable digestion product is the nucleosome core particle, which consists of the octamer and one and three-quarter turns of DNA. It and the chromatosome exploit to the fullest the flexibility of DNA. The problem of compaction is solved by bending the DNA into a supercoil with a curvature near the limit that is possible without disrupting the duplex severely. At the same time the DNA is left in a relatively accessible position on the surface of the complex.

The nucleosome is only the lowest level of chromatin structure. At the next level strings of nucleosomes appear to wind up into arrays resembling a solenoid to form fibers about 30 nanometers (millionths of a millime-

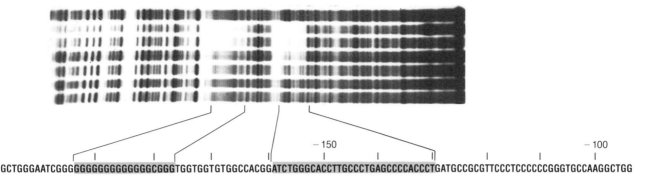

GCTGGGAATCGGGGGGGGGGGGGGGGCGGGTGGTGGTGTGGCCACGGATCTGGGCACCTTGCCCTGAGCCCCACCCTGATGCCGCGTTCCCTCCCCCCGGGTGCCAAGGCTGG

GGGCCCCTCCGGAGATGCAGCCAATTGCGGGGTGCCCGGGGAAGAGGAGGGGCCGGCGGAGCGGATAAAAGTGGGGACACAGACGG

CHROMATIN STRUCTURE is depicted near the promoter of an active gene: the adult beta globin gene in chicken red blood cells. The sequence begins at the bottom right, just upstream (toward the 5' end) from the start of transcription at position +1. In cells in which the globin gene is inactive the entire region is covered by nucleosomes, but in the red blood cells expressing the gene one nucleosome is missing. Experiments by Beverly M. Emerson in the author's laboratory show that protein factors in the cell nucleus can exclude the nucleosome from that one site by binding tightly to the DNA. An experiment to locate the binding site is shown here. When a strand of DNA radioactively labeled at one end is digested lightly with a nuclease, an enzyme that cuts DNA, the sites that bind the factor are protected from the attack. The autoradiogram (top) shows the migration of DNA fragments in a gel; they move from right to left, with the smaller fragments moving faster. The top lane of the autoradiograph gives the migration pattern of naked DNA. Beginning with the bottom lane, increasing amounts of factor were added. Two regions of protection (color) appear. The left-hand region contains a long string of GC base pairs capable of assuming an unusual secondary structure. The right-hand region contains an almost perfect inverted repeat. The gray area indicates the approximate position of the first nucleosome following the nucleosome-free zone.

ter) thick, as was originally suggested by Aaron Klug and John T. Finch of the Medical Research Council's Laboratory of Molecular Biology in Cambridge. Physicochemical studies in my laboratory and in others have produced detailed information on the packing of the chromatosomes within the fiber [see bottom illustration on opposite page]. Each solenoidal turn includes about 1,200 base pairs of DNA. One of the effects of this orderly compaction is to enable sequence elements separated by considerable distances along the DNA to come close to one another in chromatin.

How can these structures in chromatin be reconciled with the mechanisms of regulation I have described? Those mechanisms seem to require the DNA to be free of encumbrances so that it can react with the proteins controlling transcription. Recent experiments suggest that nucleosomes are disrupted in the promoter regions of genes that are being transcribed, thus making way for the regulatory proteins.

Such nucleosome-free domains reveal themselves by their unusual sensitivity to digestion by nucleases. This phenomenon was first observed in eukaryotes by Sarah C. R. Elgin and Carl Wu, who were then at Harvard University. It has been studied extensively by Harold Weintraub of the Fred Hutchinson Cancer Research Institute in Seattle. Alexander Varshavsky of the Massachusetts Institute of Technology described a similar phenomenon in the SV40 virus. In 1981

James D. McGhee, William Wood and I mapped a hypersensitive domain about 200 nucleotides long in the beta-globin gene of nuclei isolated from the red blood cells of chickens. These cells express the gene; in other kinds of chicken cells that do not express the gene no hypersensitivity to nuclease is observed.

It is obvious from the size of the hypersensitive domain in the beta-globin gene that it cannot contain a normal nucleosome. As Beverly M. Emerson in my laboratory has shown, other proteins are bound to this region, and in the test tube they prevent the binding of histones. We have purified these proteins in part and studied their interaction with DNA. There appear to be two or more distinct proteins, which bind tightly and highly specifically to well-defined sequences in the hypersensitive region.

The binding sites can be determined by means of "footprinting" experiments, in which the DNA-protein complexes are treated with nucleases. Where the protecting proteins are present, they prevent the digestion of the sequences to which they are attached. Quite recently P. David Jackson, also working in my laboratory, has developed a method for determining the footprint of bound proteins directly within the nucleus. Examination of nuclei from the chicken red blood cells reveals a pattern of DNA protection in the vicinity of the beta-globin gene quite similar to the pattern found in the binding experiments in vitro, showing that our partially purified spe-

cific factors bind to the same sites within the nucleus. There is good reason to believe these protecting proteins must help to regulate the expression of the globin gene.

The structure in the neighborhood of the active globin gene serves to illustrate the complexity of the deformations and interactions that characterize functioning DNA. The hypersensitive domain contains DNA segments that may assume partially single-stranded conformations; it also contains an inverted repeat and numerous sites of methylation. In addition to the sequences that bind protein within the hypersensitive domain, the CAAT and TATA sequences, closer to the site where transcription is initiated, probably interact with their own site-specific proteins. Our intranuclear footprinting suggests that nucleosomes lie beyond this region in each direction, forcing their characteristic distortion on the DNA.

Although I have emphasized mechanisms of gene expression, DNA also engages in complicated reactions associated with its replication and with the rearrangement of DNA segments within the genome. In all these reactions the reactivity and conformational motility of DNA play an important role. During the next decade the methods of structural and physical chemistry, combined with the powerful techniques now available for isolating and manipulating DNA sequences, are likely to provide detailed information about the mechanisms of these biochemical reactions.

# 3

# RNA

# RNA

*In all cells genetic information stored in DNA is converted into protein by RNA, which usually must be processed, even spliced, to serve its function. The first genes may have been spliced RNA*

by James E. Darnell, Jr.

In all living cells genetic information stored in DNA directs the manufacture of proteins, and in all cells the information transfer is carried out by ribonucleic acid, or RNA. The DNA message is transcribed by messenger RNA and carried to the structures called ribosomes. There the messenger is translated into specific proteins with the help of transfer RNA, which latches onto amino acids and attaches them to the growing protein chain. The ribosomes themselves contain a third type of RNA that serves as a structural component. In each of these three roles RNA is every bit as crucial to the normal functioning of a cell as DNA and proteins are; it is the link between them. Were it not for RNA, the genetic message would be inert, without means of expression.

The expression of different genes—and the resulting synthesis of different proteins—is what distinguishes, say, a brain cell from a muscle cell; generally speaking all cells in an individual organism contain the same DNA, but not all genes are active in every cell. In the 1960's the question of how gene expression is controlled led a few investigators to focus on how RNA is made, particularly messenger RNA (mRNA). At the time it was widely assumed that an mRNA molecule is simply copied from a single region of DNA. The expression of genes, according to this view, is controlled entirely by switching them on and off, that is, by determining which ones are transcribed.

For bacteria, which are simple cells without a nucleus, the assumption proved largely correct. In all eukaryotic (nucleated) cells, however, the manufacture of mRNA has turned out to be more than just a matter of transcription. The primary RNA transcript undergoes extensive processing in the nucleus before passing through nuclear pores into the cytoplasm. The processing can be quite complex. Perhaps the most startling development in molecular biology since the discovery of the DNA double helix has been the recognition that many eukaryotic genes are split, such that the primary RNA molecule must be cut and the pieces spliced to form mRNA. The messenger is not a verbatim transcript of DNA; instead, like a motion picture or a newspaper story, it has been heavily edited. In the process much extraneous material has been excised.

The discovery of RNA processing raised the possibility that gene expression might not necessarily be controlled at the level of transcription. For example, a single primary transcript might be cut or spliced in different ways to form different mRNA's coding for different proteins. A number of cellular genes have indeed been found to be controlled by such differential processing. Yet most are not; for most genes that have been investigated experimentally, the conventional view—that the expression of a gene is controlled by deciding whether to transcribe it or not—has been shown to be true. RNA processing adds a new dimension to the picture of how cells operate, but it is not the chief explanation of how the cells of an individual organism get to be different.

On the other hand, RNA processing may yield some clues as to how cells evolved in the first place. RNA was quite likely the first biopolymer; short chains of RNA, but not DNA or protein chains, form spontaneously in an environment like the one that might have prevailed on the primitive earth. Furthermore, cleavage and splicing of RNA at specific sites are now known to occur in some cases in the absence of the proteins (enzymes) that facilitate these reactions in modern cells. This suggests RNA splicing is not a recently evolved complexity. Even before cells arose it may have served its present purpose of bringing together separate pieces of useful information. I think it likely, as do many other workers, that RNA splicing contributed to the first successful gene-directed synthesis of proteins. Only later did RNA give rise, probably by reverse transcription, to DNA, which then became a secure storehouse for genetic information.

The evolutionary choice of DNA as the information-storage molecule may be reflected in one of the two important chemical differences between RNA and DNA. Both are made up of nucleotides, each of which consists of a nitrogen-containing base, a five-carbon sugar and a phosphate group. Whereas in RNA the sugar is ribose, in DNA it is deoxyribose. The difference lies at the 2' position of the sugar ring: ribose has a hydroxyl (OH) group attached to that carbon, but deoxyribose has a hydrogen atom. The 2' hydroxyl is left free when ribonucleotides are linked to form RNA, and it renders RNA chemically less stable than DNA; in an aqueous solution RNA undergoes hydrolytic cleavage at a much faster rate. As a result DNA is better suited to the function of preserv-

**RNA MIGRATES THROUGH PORES** from the nucleus, where it is transcribed from DNA, to the cytoplasm, where the genetic message is translated into proteins. In the electron micrograph, which was made by Nigel Unwin of the Stanford University School of Medicine, a small section of the nuclear envelope of a frog oocyte is enlarged about 90,000 diameters. At the edges of the pores there are ring-shaped structures, and in the center there is a plug from which eight spokes radiate. Each pore complex is about a tenth of a micrometer in diameter; the nuclear envelope shown here has some 10 million of them. Before leaving the nucleus, RNA transcripts are processed to form finished molecules. To form messenger RNA, for example, selected pieces must often be cut from the transcript and spliced.

ing information over a long period. (For the same reason RNA is harder to study in the laboratory, which explains in part why the study of RNA chemistry has until recently lagged behind the study of DNA.)

The other chemical difference between the two nucleic acids is in one of their four bases: thymine (*T*) in DNA is replaced in RNA by the closely related base uracil (*U*). Like thymine, uracil is complementary only to adenine (*A*). During RNA synthesis the adenine in DNA is transcribed to uracil and thymine is transcribed to adenine. Cytosine (*C*) is transcribed to guanine (*G*) and vice versa.

Transcription begins when an enzyme called RNA polymerase binds to a specific base sequence on DNA (the promoter). The enzyme unwinds part of the double helix, exposing two single strands of DNA, one of which is then transcribed. As the polymerase moves along the DNA, RNA nucleotides with bases complementary to those of the DNA nucleotides are added to the growing RNA chain one at a time. Each incoming nucleotide bears a triphosphate group at the 5′ position. In the synthesis reaction a pyrophosphate (a two-phosphate group) is lost, and the 5′ end of the incoming nucleotide is linked by a phosphodiester (O–P–O) bond to the 3′ hydroxyl of the nucleotide at the end of the chain. Thus an RNA molecule always grows in the 5′-to-3′ direction. The polymerase continues along the DNA chain until it passes another special base sequence called the termination signal. It thereupon releases the single strand of RNA as well as the DNA, which reforms a double helix.

The regions of DNA containing the instructions for protein synthesis are transcribed into mRNA. Each sequence of three mRNA nucleotides is a "codon" encoding one amino acid. Since a protein may consist of between 100 and 1,000 amino acids, an mRNA must be at least between 300 and 3,000 nucleotides long.

The translation of mRNA into protein at the ribosomes is a complex process. Two types of RNA that do not code for proteins themselves take part in decoding the information carried by mRNA: transfer RNA (tRNA) and ribosomal RNA (rRNA). Both are synthesized in the same manner as mRNA. Molecules of tRNA are small, about 70 to 80 nucleotides long, and they have a folded, three-dimensional structure. Each tRNA recognizes and binds, with the help of a special en-

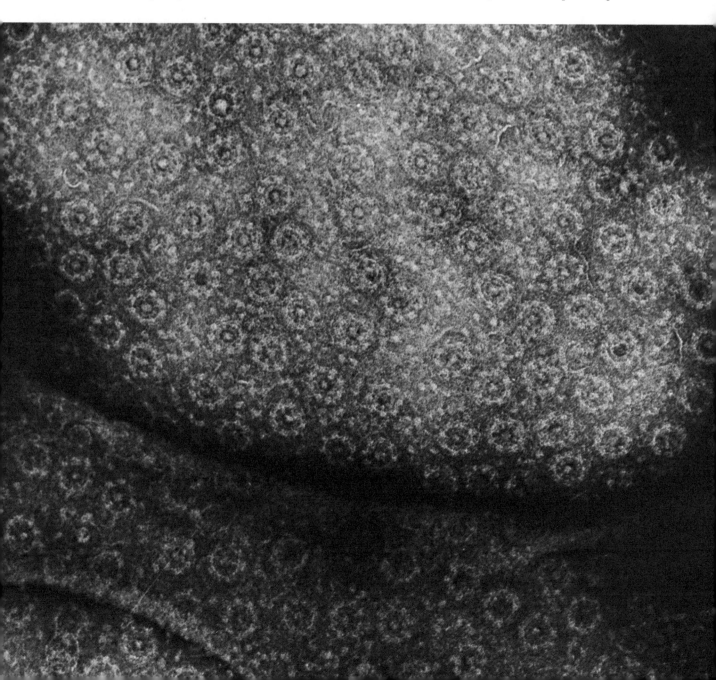

**CHEMICAL STRUCTURE OF RNA differs from that of DNA in two ways** (*dark color*). First, the DNA base thymine is replaced in RNA by uracil, which has a hydrogen rather than a methyl group attached to one of its carbons. Second, the sugar in RNA, ribose, has a hydroxyl (OH) rather than a hydrogen at its 2′ carbon. In RNA and DNA synthesis the 5′ position of an incoming nucleotide (in this case adenosine triphosphate) is joined to the 3′ hydroxyl of the last nucleotide by a phosphodiester linkage; a pyrophosphate is given off.

zyme, a single one of the 20 different amino acids found in proteins. At the opposite end of the tRNA molecule from the amino acid binding site there is a loop containing the "anticodon," a nucleotide triplet that is complementary to a specific mRNA codon.

On the surface of a ribosome tRNA's carrying amino acids are brought in contact with an mRNA molecule. A ribosome consists of a large and a small ribonucleoprotein particle, each of which is a collection of protein molecules bound to rRNA. In eukaryotes the large particle contains three rRNA molecules, a long one (about 4,500 nucleotides) and two short ones (about 160 and 120 nucleotides). The small particle contains a single long rRNA (about 1,800 nucleotides). The rRNA helps to give the ribosome its grooved structure, which enables it to accommodate an mRNA and a protein at the same time.

As the ribosome moves along the mRNA one codon at a time, tRNA's with the appropriate anticodons are selectively bound to the mRNA, and the amino acids they carry are linked to the growing protein chain. The sequence of codons on the mRNA thus dictates the amino acid sequence. The ribosome catalyzes the entire process, bringing together not only the tRNA's and the mRNA but also the specialized proteins and other factors that initiate the "reading" of mRNA, forge the peptide bonds between amino acids and release the finished protein when the genetic message has been completely translated.

Although this description of protein synthesis applies to all cells, both eukaryotes and prokaryotes (bacteria), most of the mechanisms were first revealed by studying the bacterium *Escherichia coli*. It was clear at an early stage that the regulation of mRNA synthesis (in other words, of DNA transcription) was a means by which gene expression and hence cell behavior might be controlled. A bacterium such as *E. coli* does not at all times contain every mRNA encoded by its DNA; rather, it transcribes certain genes and makes the corresponding proteins as it needs them, in response to environmental stimuli. In the early 1960's François Jacob and Jacques Monod of the Pasteur Institute in Paris proposed that the *E. coli* gene for the lactose-digesting enzyme beta-galactosidase is controlled by a repressor: a regulatory protein that blocks the transcription of the gene when the enzyme is not needed by forming a bond with the DNA near the start of the coding sequence. The Jacob-Monod model was subsequently found to apply generally to the control of gene expression in bacteria. Many regulatory proteins have now been found that either decrease or increase the rate of transcription of bacterial genes by binding to specific DNA sites.

In multicellular organisms each cell is part of an interdependent set and is not forced to respond constantly to changes in the external environment, as bacteria are. Instead the most dramatic variations in cellular behavior are the differences within the same organism among cell types, all of which contain the same DNA. For example, in human beings only red blood cells make hemoglobin, which carries oxygen throughout the body. Similarly, a liver cell makes at least 50 proteins that are either not made at all in other cells or made at much lower rates.

As soon as Jacob and Monod announced their model it was tempting to believe that such cell specialization, like the variable behavior of bacteria, is controlled at the level of gene transcription. It has taken 20 years, however, to prove the surmise correct. One reason is the discovery that most RNA is processed after transcription, which raised the possibility that gene expression might be controlled at the processing stage as well. For example, a particular RNA transcript might be synthesized in all cells but processed successfully only in some; or the processing rate might vary, so that different cell types would contain different amounts of a particular mRNA and its corresponding protein. In the late 1960's the question of how mRNA is made became a key part of the larger question of how genes are controlled.

The first evidence for RNA processing had involved rRNA rather than mRNA. In 1960, when I was at the Massachusetts Institute of Technology, my colleagues and I began to analyze newly formed nuclear RNA in human cells. (New RNA is identified by allowing the cells briefly to synthesize RNA from radioactively labeled nucleosides, the precursors of nucleotides.) To our surprise we found no labeled nuclear RNA corresponding in size to the rRNA in mature ribosomes.

EUKARYOTE

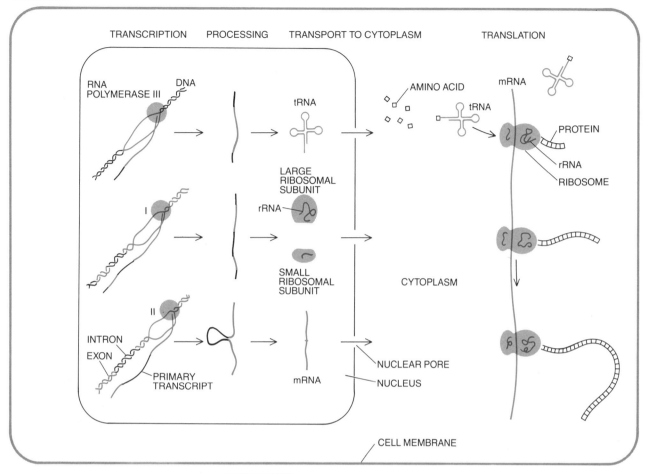

TRANSCRIPTION    PROCESSING    TRANSPORT TO CYTOPLASM    TRANSLATION

RNA POLYMERASE III    DNA

tRNA

AMINO ACID    mRNA

tRNA

PROTEIN

rRNA

RIBOSOME

LARGE RIBOSOMAL SUBUNIT

rRNA

I

SMALL RIBOSOMAL SUBUNIT

CYTOPLASM

II

INTRON

EXON

PRIMARY TRANSCRIPT

mRNA

NUCLEAR PORE

NUCLEUS

CELL MEMBRANE

PROKARYOTE

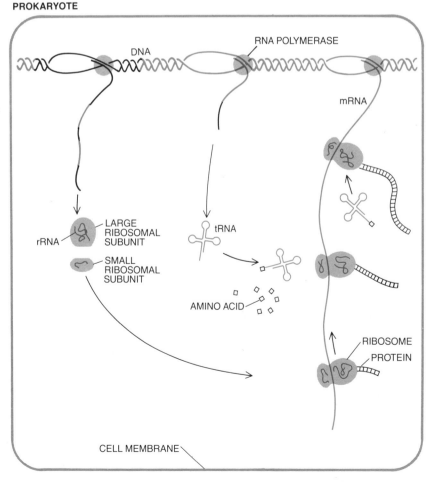

DNA    RNA POLYMERASE

mRNA

rRNA    LARGE RIBOSOMAL SUBUNIT

SMALL RIBOSOMAL SUBUNIT

tRNA

AMINO ACID

RIBOSOME

PROTEIN

CELL MEMBRANE

**PROTEIN SYNTHESIS** involves the same three types of RNA in eukaryotic and in prokaryotic cells, but with an important difference: in eukaryotes the protein-coding sequences of DNA (exons) are often separated by intervening sequences (introns) that must be excised from a primary transcript to make messenger RNA (mRNA). In both kinds of cell, transfer RNA (tRNA), ribosomal RNA (rRNA) and mRNA are made by the transcription of one strand of the DNA double helix. Three different polymerase enzymes catalyze these reactions in eukaryotes, whereas in prokaryotes there is only a single type of polymerase. In both kinds of cell the tRNA and rRNA primary transcripts must be processed. The ends of the tRNA transcript are cut and the molecule assumes a looped structure. A single rRNA transcript is cut in several places to form two major types of rRNA, which are then bound to protein molecules to form ribosomal subunits. In prokaryotes, which have no nucleus, the mRNA transcript is generally not processed; ribosomes and tRNA's carrying amino acids begin translating the mRNA into a sequence of amino acids—a protein—as it is being made. In eukaryotes the nuclear envelope probably facilitates the removal of introns and splicing of exons from the primary mRNA transcript by protecting it from immediate translation. The mRNA is "read" only after it (like tRNA and the ribosomal subunits) has exited the nucleus through pores in the envelope. Only in the cytoplasm are the mature mRNA, tRNA and ribosomes united.

Instead we were able to show that both long and short rRNA's are produced from a single longer precursor molecule—a primary transcript—by cutting the precursor at specific sites. In the ensuing years the same type of processing was observed in many other types of cell; it was observed in the production of both rRNA and tRNA, individual molecules of which are also made by shortening a primary transcript. Soon it became clear that all rRNA's and tRNA's, in eukaryotes as well as in prokaryotes, must both be shortened and receive various chemical additions before they are ready to take part in the translation of mRNA.

What about the mRNA itself? In bacteria, which have no nucleus, it seemed unlikely that mRNA could be processed, because it is engaged by waiting ribosomes and tRNA's even before its synthesis has been completed. In eukaryotes, on the other hand, mRNA is made in the nucleus and must then migrate to the cytoplasm to be translated into protein. Between transcription and translation there is thus ample opportunity for processing.

To determine whether such processing occurs workers had first to confront a daunting problem: a typical eukaryotic cell contains thousands of kinds of mRNA. The solution was to study animal cells infected by viruses. Viruses whose genome consists of DNA usurp both the protein- and the mRNA-making machinery of the host cell, but their DNA encodes far fewer proteins than the host's DNA and therefore is transcribed into far fewer mRNA's. Consequently the manufacture of viral mRNA can be treated as a simple model for the way an animal cell makes its own mRNA. By this approach the basic steps in the manufacture of eukaryotic mRNA were revealed between 1970 and 1977.

The complexity of the process compared with the one in bacteria is evident from the first step. Whereas in bacteria a single polymerase enzyme catalyzes all RNA synthesis, in eukaryotes there are three distinct polymerases: type I makes the precursor of rRNA's, type II makes the mRNA precursor and type III makes the precursor of tRNA's and of other small RNA's. The synthesis of an mRNA begins when an RNA polymerase II somehow recognizes and binds to a "promoter" region on a DNA double helix. The roles of the various nucleotide sequences in the promoter region are not fully understood, but the sequence *TATA* is known to be involved in positioning the polymerase correctly. Transcription begins between 20 and 30 nucleotides past the *TATA* site.

Within about a second, before the RNA is more than 30 nucleotides long, a chemically protective "cap" is added to the nucleotide at the beginning of the chain (the 5' end). The cap, which is retained in all eukaryotic mRNA, consists of a methylated guanosine (a nucleoside incorporating the base guanine) linked to the first nucleotide by a triphosphate bridge. After the cap is attached the polymerase continues to add nucleotides to the 3' end of the chain at the rate of from 30 to 50 per second. The resulting primary transcript can be as many as 200,000 nucleotides long, but the average in human cells is about 5,000 nucleotides.

Processing of the completed transcript begins with the addition of a "tail" to a nucleotide that will become the terminal (3') end of the mRNA. In 1970 and 1971 workers in several laboratories found that most mammalian mRNA's have grafted on their 3' end a series of between 150 and 200 adenine nucleotides, called poly(*A*) for short. It is now known that the primary transcript is cut and poly(*A*) is added to the free end about 20 nucleotides past the sequence *AAUAAA*. These reactions occur within a minute after the polymerase has crossed the poly(*A*) site.

After the cap and the tail had been recognized on finished mRNA in the

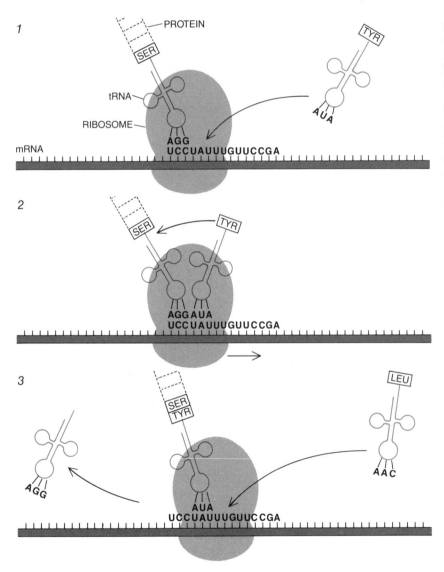

**TRANSLATION of mRNA into protein at a ribosome follows the same steps in both eukaryotes and prokaryotes. Each nucleotide triplet, or codon, on the mRNA chain encodes a specific amino acid. Each molecule of tRNA in turn binds only the amino acid corresponding to a particular codon. A tRNA recognizes a codon by means of a complementary nucleotide sequence called an anticodon. Here the addition of one amino acid to a protein chain is shown. An incoming tRNA molecule carrying the amino acid tyrosine binds to the codon exposed at a binding site on the ribosome (*1*). The tyrosine forms a peptide bond with serine, the last amino acid on the protein chain (*2*). As the ribosome advances one codon (*3*), exposing the binding site to the next incoming tRNA, the serine tRNA is released.**

cytoplasm, investigators began to look for the same structures on molecules in the nucleus in the hope of identifying the precursors of mRNA. By 1976 such precursors had been found among newly formed nuclear RNA, but a puzzle remained. The nuclear molecules were on the average about 5,000 nucleotides long, whereas the average length of the mRNA's in the cytoplasm was only about 1,000 nucleotides. Clearly the precursor had to be cut to form mRNA, but how were the cap and the tail preserved?

In retrospect the answer seems obvious: parts of the transcript must be cut out of the middle, and the ends must then be spliced. RNA splicing was first observed in the case of the adenovirus, which infects the human upper respiratory tract. My colleagues and I had found that the mRNA's for a certain group of adenovirus proteins (called late proteins because they are synthesized at a late stage of the infection) are all derived from a single primary transcript. In 1977 a group at M.I.T. led by Phillip A. Sharp and a group at the Cold Spring Harbor Laboratory (including Richard J. Roberts, Thomas R. Broker, Louise T. Chow, Richard E. Gelinas and Daniel Klessig) used the technique of molecular hybridization to show how the derivation occurs.

The technique relies on the fact that an mRNA will react by complementary base-pairing with the DNA from which it was transcribed, forming a double-strand hybrid. (One DNA strand is displaced.) The hybrid can be visualized in an electron micrograph. When this experiment was done with the nucleic acids coding for adenovirus late proteins, each mRNA hybridized with four regions on the DNA; the four regions were separated by long stretches of DNA that looped out from the hybrid, evidently because their nucleotide sequence had no complement on the mRNA. The M.I.T. and Cold Spring Harbor groups concluded that the mRNA must be formed by splicing four pieces from distant regions of the primary transcript. In the process the intervening sequences are eliminated, and so the finished mRNA is much shorter than the primary transcript. The 5' cap and 3' poly(A) tail, however, are preserved.

Splicing of the adenovirus primary transcript takes between 20 and 30 minutes, but in other cases the operation takes as little as five minutes. Following the lead of the M.I.T. and Cold Spring Harbor workers other investigators soon showed that many mRNA's, both viral and cellular, are spliced from pieces of a long precursor, with other pieces being discarded. Later it became possible to find specif-

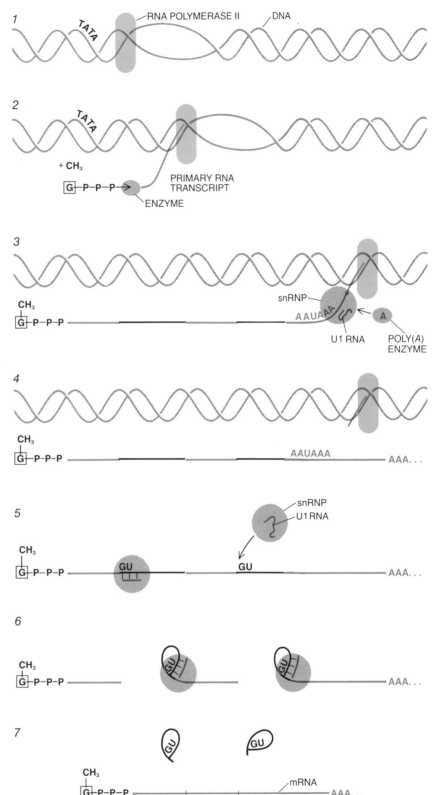

**MANUFACTURE OF MESSENGER RNA** in a eukaryote begins when transcription of a DNA is initiated by an RNA polymerase some 20 to 30 nucleotides "downstream" from the sequence *TATA* (*1*). Before the transcript is more than 30 nucleotides long its 5' end is "capped" with a guanosine that is attached by a triphosphate group and methylated (*2*). The polymerase moves along the DNA (*3*), transcribing both exons (*color*) and introns (*black*). Eventually it transcribes the sequence *AAUAAA*. About 20 nucleotides downstream the transcript is cut; the reaction probably involves a small nuclear ribonucleoprotein (snRNP) containing a uracil-rich "U1 RNA." An enzyme adds a "tail" of 150 to 200 *A*'s to the cut 3' end (*4*). The transcript is cut at the introns (*5*), probably with the help of snRNP's; U1 RNA binds to a complementary sequence (including *GU*) at the beginning of introns. The introns are folded into lariats (*6*) and excised, and the exons are spliced to form mRNA (*7*).

ic splicing sites by analyzing the nucleotide sequence of a DNA and the corresponding mature mRNA. It is now generally accepted that many eukaryotic genes are in pieces and cannot be expressed without extensive processing of the RNA transcript. Walter Gilbert of Harvard University has coined the term exons to refer to the pieces of a gene that are actually expressed in mRNA and protein; the nucleotide sequences intervening between the exons, the ones subsequently spliced out of the RNA, are called introns.

Given that an error of just one nucleotide might render the protein worthless, what mechanisms ensure that the exons are joined correctly? Current research is revealing the biochemistry of splicing in considerable detail. Splicing has now been reproduced in the laboratory: two exons separated by an intron on experimental

RNA molecules have been joined by exposing the molecules to cell extracts. Although by no means all of the enzymatic steps in this process are understood, some of the crucial elements in the extracts have been identified.

One of these elements is a class of ribonucleoprotein particles found in the nuclei of all animal cells. Known as small nuclear ribonucleoproteins (snRNP's), they each consist of a group of protein molecules bound to a single molecule of RNA. The RNA, which is different from the three main types described above, is between 100 and 300 nucleotides long. There are at least 10 distinct types of these short RNA molecules. Six of them are notably rich in the base uracil and so are called U-RNA's; the one referred to as U1 RNA is of particular interest.

The evidence that snRNP's play a crucial role in splicing comes from experiments involving antibodies to their

proteins. For reasons that are not understood, such antibodies are made in profusion by people suffering from certain autoimmune diseases, including systemic lupus erythematosus. Joan A. Steitz and her colleagues at Yale University have isolated the antibodies to snRNP's and, together with Sharp's group at M.I.T., have added them to cell extracts known to be capable of splicing RNA. When the antibodies are added to an extract before the unspliced RNA transcripts, splicing does not take place. Furthermore, neither does polyadenylation. These observations indicate that snRNP's participate in cutting the primary transcript at the poly($A$) site as well as in locating and splicing out introns. It is not known how the snRNP's carry out the tasks or whether the same particles take part in both polyadenylation and splicing. One popular hypothesis holds that a single large, ribosomelike structure made up of snRNP's and other particles engages in both processes. In any event, splicing almost certainly involves the recognition of an intron by an snRNP containing U1 RNA, which binds to complementary bases at the beginning of the intron.

The discovery of introns in 1977 came as a complete surprise; until then molecular biologists had always assumed that a gene was a single, continuous stretch of DNA encoding a single protein. In addition to the issue of how splicing is accomplished, the discovery immediately raised two broader questions. First, how important is RNA processing, and splicing in particular, in the regulation of eukaryotic gene expression? Second, have genes always been split or were introns gradually introduced over the course of evolution?

Once it was known that genes are in pieces one could envision a central role for RNA processing in regulating gene expression. Not only might the amount of a particular protein in a cell be determined by the amount of successfully processed mRNA, but also a single gene might encode more than one protein, and the decision of which one to make at a given time or in a given cell would be a processing decision. A primary transcript with two or more poly($A$) sites could be cut on some occasions at one site and on other occasions at another site, producing mRNA's with different 3' ends and therefore proteins with different terminals. A transcript containing three or more exons could be spliced in such a way that all the exons are included in the mRNA or as few as two. Only two constraints seemed likely to limit the splicing choice: the end exons, to

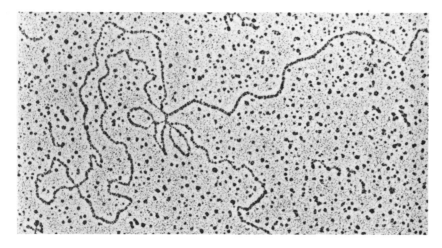

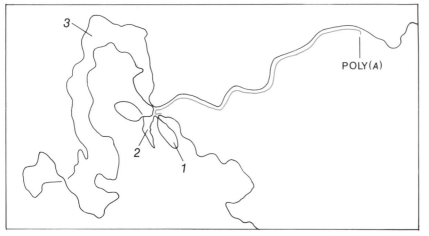

**SPLICING OF MESSENGER RNA** was discovered in studies of the adenovirus genome. The electron micrograph shows the result of an experiment in which the mRNA (*colored line on map*) for an adenovirus protein was allowed to hybridize with a single strand of the DNA for that protein (*black*). Where their nucleotide sequences are complementary the two molecules form a double-strand hybrid. Three regions of DNA for which there are no complementary sequences on the RNA form loops (*1–3*). Evidently the RNA transcripts of these regions have been excised from the mRNA, which has then been spliced. The electron micrograph was made by Louise T. Chow, then at the Cold Spring Harbor Laboratory.

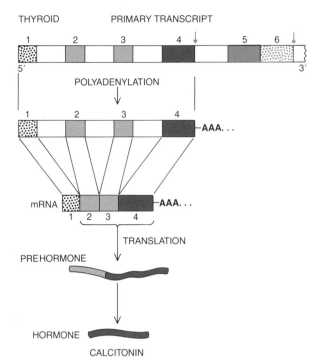

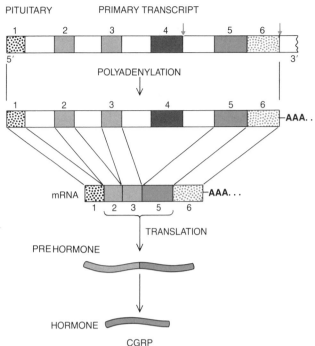

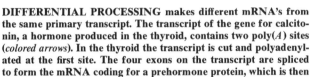

**DIFFERENTIAL PROCESSING** makes different mRNA's from the same primary transcript. The transcript of the gene for calcitonin, a hormone produced in the thyroid, contains two poly($A$) sites (*colored arrows*). In the thyroid the transcript is cut and polyadenylated at the first site. The four exons on the transcript are spliced to form the mRNA coding for a prehormone protein, which is then enzymatically cleaved to form calcitonin. In the pituitary gland and in some nerve cells the same primary transcript is cut instead at the second poly($A$) site. At the splicing stage the coding sequence for calcitonin (*dark gray*) is excised along with the introns; in its place a sequence coding for a hormone called calcitonin-gene-related protein (*color*), or CGRP, is incorporated in the finished mRNA.

which the 5′ cap and the 3′ tail are attached, must be preserved, and the exons must remain in the same order.

So far several dozen cellular and viral genes have been found to encode more than one mRNA. Such genes are called complex transcription units. A good example is the late-protein transcription unit of the adenovirus, which contains five poly($A$) sites and numerous exons. Its primary transcript can be spliced to yield a dozen different mRNA's. Each mRNA consists of four exons spliced together. The first three exons are common to all the messengers; the last exon contains the coding sequence for a particular protein.

In animal cells too a number of complex transcription units are now known to produce different mRNA's under different circumstances. Some of these units have been found in the muscle cells of mammals and insects. The cells in muscles with different functions require related but slightly different proteins. In a given organism such proteins often seem to be encoded by a single gene whose primary transcript can be spliced or polyadenylated in various ways.

A particularly striking example of this phenomenon—the control of gene expression by RNA-processing decisions—involves a complex transcription unit isolated from rat DNA by Ronald Evans and Michael G. Rosen-

feld of the Salk Institute for Biological Studies. The DNA segment in question encodes calcitonin, a hormone produced in the thyroid gland. Evans and Rosenfeld observed, however, that it also hybridizes with an mRNA made in the pituitary gland (and later found in certain nerve cells). The workers showed that the calcitonin primary transcript is present in pituitary cells as well as in the thyroid. The transcript contains two poly($A$) sites. In the thyroid, cleavage at the first site followed by exon splicing yields calcitonin mRNA; in the pituitary the transcript is cleaved at the second poly($A$) site and the exon containing the calcitonin coding sequence is spliced out. The result is an mRNA coding for a previously unknown neuropeptide, now called calcitonin-gene-related protein, or CGRP.

If the calcitonin gene is a good example of gene regulation through RNA processing, it also illustrates a more fundamental truth, namely that the dominant mechanism of gene control in eukaryotes, as in bacteria, is not differential processing but differential transcription. Most rat cells manufacture neither calcitonin nor CGRP, and in such cells the calcitonin gene is not transcribed in the first place. To molecular biologists this comes as no surprise; even after RNA processing was

discovered most investigators still expected transcriptional control to be dominant. The alternative—transcribing all genes in all cells but processing the transcripts differently—was always assumed to be unlikely. Yet only recently, with the advent of gene-cloning techniques, has the validity of the assumption been fully demonstrated.

My colleagues and I have done cloning experiments whose results confirm that unexpressed genes are generally not transcribed. First we extracted all the mRNA from mouse liver cells and copied it into DNA using reverse transcriptase, an enzyme that catalyzes the reverse transcription of RNA into DNA. The DNA was inserted into plasmids (small circles of bacterial DNA) and introduced into *E. coli,* which were grown in culture. The cloning process yielded experimentally useful amounts of individual mouse genes, because the bacteria in a given colony all contained the same mouse-DNA insert. By exposing the DNA to total mRNA from various types of cell, we were able to identify genes encoding liver-specific proteins (such DNA's hybridized only with mRNA from liver cells) and genes encoding common proteins (which hybridized with mRNA from all types of cell).

Obviously the mRNA's for liver-specific proteins were not made in cells outside the liver, but was it because the transcripts were not processed properly in such cells or because the genes were not transcribed at all? To select between these alternatives we collected nuclei from brain, kidney and liver cells and allowed them to synthesize RNA from radioactively labeled precursors for a brief period (about 10 minutes). When we removed RNA transcripts from the nuclei, we knew that any labeled transcripts had to be newly formed and therefore unprocessed. If the liver-specific genes were transcribed in brain and kidney nuclei, the labeled RNA from those nuclei should have contained liver-specific transcripts. When the extracts were exposed to the various cloned DNA's, however, the RNA from brain and kidney nuclei hybridized only with the DNA's encoding common proteins and not with the liver-specific ones. Clearly the genes for the 12 liver proteins we examined were not being transcribed in the brain and the kidney.

On the basis of many similar experiments the result can be stated more generally: In multicellular organisms most tissue-specific decisions to produce certain proteins and not others are made at the transcriptional level. How exactly such choices are made is a question currently being investigated in dozens of laboratories. So far no conclusive answers have been found.

Although RNA processing is not a major basis of cell differentiation, some form of it takes place in all cells, and in most cases it is essential to their function. When did it evolve? Splicing of mRNA primary transcripts has now been observed in bacteria, but much less often than in eukaryotes; bacteria contain far fewer introns. Until recently most biologists agreed that bacteria, by virtue of their simplicity, must be closely related to the earliest cells, and that eukaryotes evolved from these ancestral prokaryotes. Consequently many workers at first thought of introns as complexities introduced relatively late in evolution.

Two separate lines of evidence undermine this view. First, there is good reason to question the belief that bacteria are evolutionarily older than eukaryotes. Second, experimental evidence suggests that RNA processing, including splicing, can occur without the proteins that catalyze the reactions in modern cells; indeed, the reactions could have occurred in a precellular environment.

The evidence against viewing bacteria as the oldest organisms comes largely from the work of Carl R. Woese and his colleagues at the University of Illinois at Urbana-Champaign. As a way of charting cell lineages Woese's group decided to compare the nucleotide sequence of ribosomal RNA from many organisms. (Ribosomal RNA was chosen because ribosomes are present in essentially similar form in all organisms; they are therefore thought to be primordial cell structures.) To their surprise the workers found that a small group of bacteria with unusual metabolisms did not fit into the lineage chart for normal bacteria. This group, called the archaebacteria, seemed to be no more closely related to normal bacteria (the eubacteria) than they were to eukaryotes. On the basis of rRNA sequences the eukaryotes also formed a distinct group

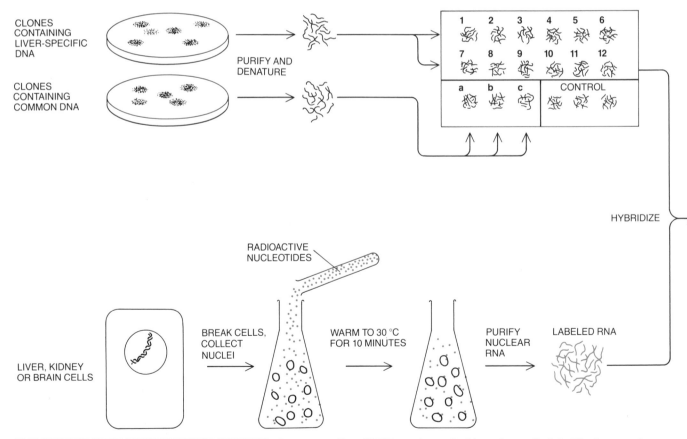

**DOMINANCE OF TRANSCRIPTIONAL CONTROL** of gene expression over control at the RNA-processing level was demonstrated with the help of gene-cloning techniques. Mouse DNA coding for liver-specific proteins was cloned in bacterial colonies, as was DNA encoding proteins common to most mouse cells. DNA's coding for 12 different liver proteins (*1–12*) and three different common proteins (*a–c*) were purified, denatured (separated into single strands) and transferred to filter paper. To determine whether all these DNA's are transcribed in various cells, hybridization experiments were done with newly formed, unprocessed RNA transcripts from liver, kidney and brain cells. New transcripts were isolated by collecting chilled cell nuclei and warming them briefly in the presence of radioactive nucleotides. Since transcription does not occur in chilled nuclei, any radioactively labeled RNA purified from the nuclei had to have been synthesized during the 10-minute warming period. When labeled RNA was exposed to the filter paper,

spanning a vast range of organisms from yeast to human beings.

Because the three groups share the same genetic code, Woese concluded they must represent separate lines of descent from a common ancestor, which he called the progenote. The eubacteria and the archaebacteria may have evolved directly from the progenote. Modern eukaryotes, on the other hand, are almost certainly the product of the fusion of an ancestral eukaryote with two types of eubacteria. The ancestral eukaryote gave rise to the modern nucleus; bacteria gave rise to mitochondria, the energy-converting organelles found in all eukaryotes, and to chloroplasts, the photosynthesizing organelles found in plants. (Both mitochondria and chloroplasts contain their own DNA and RNA; moreover, the rRNA sequences of mitochondria are similar to those of purple sulfur bacteria, whereas the rRNA of chloroplasts is closely related to that of the cyanobacteria.) According to Woese, the eukaryotic nucleus is as old as bacteria. Since the nucleus is where RNA processing takes place, there is no reason to assume that processing began relatively late.

It may have begun even before the progenote emerged. For years many investigators have proposed that RNA must have been the first biopolymer: like DNA it can store information (although it is not as good at storing it securely), and unlike DNA it is essential in the decoding of stored information. Recent studies of RNA chemistry strongly support the idea that RNA came first. Leslie E. Orgel and his colleagues at the Salk Institute have shown that the synthesis of RNA oligonucleotides proceeds spontaneously in an environment marked by high salt and nucleotide concentrations; such an environment could have prevailed on the primitive earth. RNA copies of RNA chains can also form spontaneously through the pairing of complementary bases. Both reactions are slow, but they both proceed in the absence of enzymes or any other proteins. It therefore seems a distinct possibility that the primordial soup contained short chains of RNA as well as individual nucleotides and amino acids. Neither DNA nor proteins polymerize nearly as readily as RNA.

The essential RNA-processing reactions—site-specific cleavage and splicing—can also occur in the absence of proteins. Sidney Altman and his colleagues at Yale have observed an example of the former process in *E. coli*. Ordinarily a tRNA precursor in the bacterium is cut by an enzyme that itself contains a small RNA molecule. Altman's group has found, however, that the small RNA alone can cleave the precursor correctly, without the help of the enzyme protein.

Perhaps even more surprising is the occurrence of protein-free RNA splicing. Thomas R. Cech and his colleagues at the University of Colorado at Boulder first documented the phenomenon in the protozoan *Tetrahymena pyriformis,* which is able to dispense with proteins in excising an intron from its rRNA transcript and in joining the two exons. Protein-free splicing has also been observed in the case of mitochondrial rRNA from the fungus *Neurospora crassa.* The nucleotide sequences of the two self-splicing rRNA's are notably similar.

Assuming that the first information-storage molecules, or genes, were molecules of RNA, the splicing and cutting of RNA may have been significant processes in precellular evolution. How RNA ever came to encode sequences of amino acids remains a mystery. It seems likely, however, that the useful coding information in spontaneously generated RNA chains would not have been in the form of long, uninterrupted nucleotide sequences. Instead short stretches of RNA encoding primitive but functional peptide chains would probably have been separated by sequences containing no coding information, which are now recognized as introns.

The hypothesis is supported by the positions of exons and introns in modern genes. In many genes, for example those encoding hemoglobin, antibodies (immunoglobulins) and certain enzymes, each exon encodes a domain of the protein molecule that is recognizable as a functional unit. The random introduction of introns over the course of evolution does not seem a likely source of such a nonrandom division of genes. It is much more plausible to assume the introns were there from the beginning. In that case the self-splicing of RNA may have been exploited even before cells evolved to remove introns and unite peptide-coding regions that together encoded larger, more useful proteins.

How did DNA evolve? Reverse transcriptase was discovered in virus-

LIVER

KIDNEY

BRAIN

WASH

**transcripts from kidney and brain cells hybridized only with the common DNA's and not with the liver-specific ones. This proved that the liver-specific genes are not transcribed in the kidney and the brain; if the genes were transcribed, with the transcripts then being destroyed at the processing stage, some liver-specific transcripts should have been present in the mixtures of unprocessed RNA from kidney and brain cells and should have hybridized with the liver-specific DNA. As was expected, transcripts from liver cells did hybridize with both types of DNA. Hybrids stay fixed to the filter paper when it is washed; since they are radioactive, they form dark spots on a photographic emulsion exposed to the paper.**

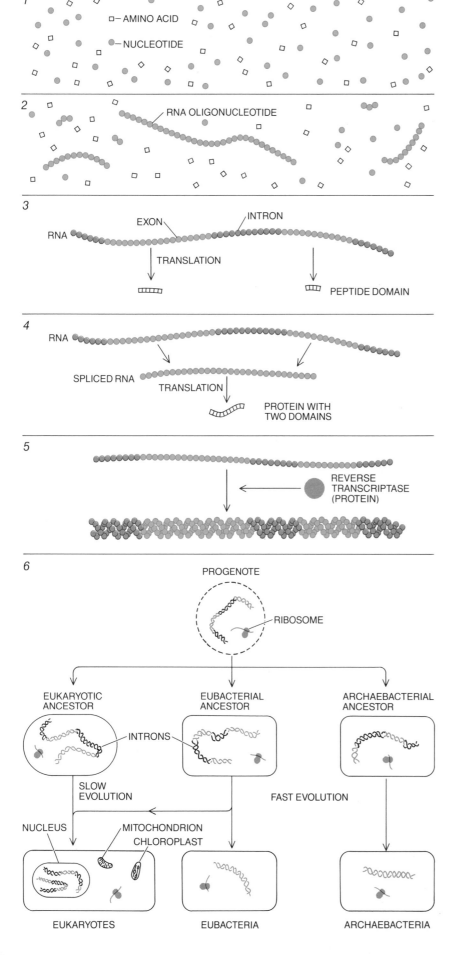

**1**

□ — AMINO ACID

● — NUCLEOTIDE

**2**

RNA OLIGONUCLEOTIDE

**3**

RNA

EXON    INTRON

TRANSLATION

PEPTIDE DOMAIN

**4**

RNA

SPLICED RNA

TRANSLATION

PROTEIN WITH
TWO DOMAINS

**5**

REVERSE
TRANSCRIPTASE
(PROTEIN)

**6**

PROGENOTE

RIBOSOME

EUKARYOTIC
ANCESTOR

EUBACTERIAL
ANCESTOR

ARCHAEBACTERIAL
ANCESTOR

INTRONS

SLOW
EVOLUTION

FAST EVOLUTION

NUCLEUS    MITOCHONDRION
CHLOROPLAST

EUKARYOTES    EUBACTERIA    ARCHAEBACTERIA

es in 1970 by Howard M. Temin of the University of Wisconsin at Madison and David Baltimore of M.I.T. More recently indirect evidence of the presence of a similar enzyme has been found in all eukaryotic cells, which suggests that enzymatically catalyzed reverse transcription of RNA into DNA is quite old. The process may therefore account for the evolutionary transfer of the information-storage function from RNA to the more stable DNA, a change that was a precondition for the emergence of self-replicating cells. Through reverse transcription the distribution of exons and introns in RNA would have been preserved in DNA.

Whether the progenote, the ancestor of all cells, was a cell itself or some intermediate form remains unclear. Certainly it must have contained DNA, probably littered with introns. The genes of the earliest eubacteria and archaebacteria most likely also contained a substantial number of introns. Through countless generations of evolution bacteria have now virtually eliminated noncoding sequences from their DNA; they have evolved into streamlined protein-making machines. The DNA of eukaryotes, including human beings, has evolved more slowly. With our cluttered genes and our reliance on RNA splicing to make sense of them, we may actually be closer than the bacteria to the primitive ancestor of all living things.

**PRECELLULAR EVOLUTION may have required the splicing of RNA. The primordial soup probably contained both nucleotides and amino acids (1). Under primordial conditions RNA chains of 30 or more nucleotides could have formed through spontaneous synthesis and copying of shorter chains (2). Somehow RNA sequences came to encode sequences of amino acids constituting peptide domains (3), but the coding regions, or exons (color), may have been interrupted by introns (gray). RNA splicing, which can occur even in the absence of proteins, may have arisen as a way of uniting exons that together encoded a protein more useful than the individual peptide domains (4). An early protein similar to the modern enzyme reverse transcriptase may, by copying RNA into DNA, have effected the changeover from an RNA-based genetic system to one based on DNA (5). The common ancestor of all cells, the progenote, contained intron-rich DNA, as well as ribosomes to translate genetic information into proteins (6) according to the now universal code. Modern eukaryotes resulted from a fusion of an ancestral nucleus with eubacteria, from which the eukaryotes obtained their mitochondria and chloroplasts. According to this view, bacterial genomes are relatively intron-free because bacteria have evolved through many more generations than the eukaryotes have.**

# 4

# PROTEINS

# Proteins

*Proteins are the molecules encoded by genes. The proteins in turn give rise to structure and, by virtue of their selective binding to other molecules, make genes and all the other machinery of life*

by Russell F. Doolittle

If DNA is the blueprint of life, then proteins are the bricks and mortar. Indeed, they serve also as the jigs and tools needed in the assembly of a cell or an organism, and they even play the role of the builders who carry out the work of assembly. Your genes supply the information, but you *are* your proteins.

Like DNA, a protein is a linear polymer: a chain of subunits linked in a continuous sequence. In other respects, however, the two kinds of molecule are quite different. Roughly speaking, all DNA molecules are alike in overall structure, and they all have the same function (that of a genetic archive). Proteins, in contrast, fold up into a remarkable diversity of three-dimensional forms, which give them a corresponding variety of functions. They serve as structural components, as messengers and the receptors of messengers, as markers of individual identity and as weapons that attack cells bearing foreign markers. Some proteins bind to DNA and thereby regulate the expression of genes; others take part in the replication, transcription and translation of genetic information. Perhaps the most important proteins are the enzymes, the catalysts that determine the pace and the course of all biochemistry.

In the study of proteins a major aim has been to decipher their structure and so learn how they work. A complete structural analysis is a laborious undertaking, and up to now biochemists have gained a thorough understanding of only a small fraction of the known proteins. Nevertheless, some general principles have emerged; substructures that are common to diverse proteins, and that probably have similar functions in many of them, can now be recognized. Of equal interest is the question of how the thousands of proteins in a typical organism have evolved and diversified. The presence of shared substructures implies a complex evolution. It is not simply a matter of one protein's being modified and thus giving rise to another; rather, fragments of genetic information must somehow be exchanged and then expressed in many proteins.

Through all the functional diversity of proteins there runs a common thread: for the most part, proteins work by selectively binding to molecules. In the case of a structural protein the binding often links identical molecules, so that many copies of the same protein aggregate to form a larger-scale structure such as a fiber, a sheet or a tubule. Other proteins have an affinity for a molecule different from themselves. Antibodies, for example, bind to specific antigens; hemoglobin binds to oxygen in the lungs and then releases it in distant tissues; regulators of genetic expression bind to specific patterns of nucleotide bases in DNA. Receptor proteins embedded in the cell membrane recognize messenger molecules (such as hormones and neurotransmitters), which may themselves be proteins that have a specific affinity for the receptors. Virtually all the activities of proteins can be understood in terms of such selective chemical binding.

The binding of a protein to the molecule it recognizes is not fixed or permanent. It is governed by a dynamic equilibrium, in which molecules are continually being bound and released. At any instant the percentage of bound molecules depends on the relative amounts of the two substances present and on the strength of the association between them. The binding strength depends in turn on how well the molecules fit together geometrically and on specific local interactions, such as electrostatic attraction or repulsion between charged regions.

Enzymes, in this respect, are much like other proteins. An enzyme recognizes a specific molecule (called the substrate) and binds to it in dynamic equilibrium; what distinguishes an enzyme is that it can bring about some chemical change in the bound substrate. The change generally entails the forming or breaking of a covalent chemical bond: the substrate may be split into two pieces, a chemical group may be added or the pattern of the bonds in the substrate may simply be rearranged.

The mechanism of enzyme action can be viewed as having three stages. First the enzyme binds to the substrate, then the chemical reaction takes place and finally the altered substrate is released. All three steps are reversible. If an enzyme binds to molecule $X$ and

**PROTEIN BINDING SITE is emblematic of the principal mechanism by which proteins do the work of biochemistry: by forming a close but generally short-lived association with another molecule. The protein is alcohol dehydrogenase, an enzyme in the liver that converts ethyl alcohol into acetaldehyde. Carbon atoms in the protein structure are white, oxygen atoms red and nitrogen atoms blue. The atoms shown in purple make up a molecule of nicotinamide adenine dinucleotide (NAD), a coenzyme that takes part in the catalyzed reaction by receiving a hydrogen ion removed from an alcohol molecule. (The alcohol is bound to a site elsewhere on the protein.) The NAD molecule fits precisely into a cleft on the protein surface and is held there by electrostatic attraction. Many proteins that bind to NAD and related coenzymes include a domain of similar structure. It is called the mononucleotide fold and may be one of the most ancient structural units in the evolution of proteins. This computer-generated image and the ones on pages 42 and 43 were made by Jane M. Burridge of the U.K. Scientific Centre of the International Business Machines Corporation.**

converts it into molecule $Y$, the same enzyme can also bind to $Y$ and change it back into $X$. Indeed, there are many possible reaction paths. A molecule of either $X$ or $Y$ could be bound but released before any change took place, or a molecule of $X$ could be converted into $Y$ and then changed back into $X$ before it was released, and so on.

It should be emphasized that the enzyme itself does not determine the direction of the reaction. The proportion of $X$ and $Y$ at equilibrium depends on thermodynamic considerations; the favored proportion is the one that minimizes the quantity called free energy. (Roughly speaking, the free energy of a system is equal to its energy minus its entropy, or disorder.) The enzyme merely hastens the attainment of equilibrium. Nevertheless, an enzyme can effectively control the course of a biochemical process. In the absence of an enzyme most biochemical reactions are extremely sluggish; the appropriate enzyme can speed them up by a factor of a million or more. Although the enzyme has no influence on whether more $X$ is converted into $Y$ or vice versa, it determines whether or not the conversion takes place at all.

An enzyme speeds a reaction by lowering an energy barrier. Even when a reaction is thermodynamically favorable—when the products have a lower free energy than the reactants—there may be an intermediate state with a higher free energy. The enzyme tends to smooth this hump in the reaction path. The mechanism varies from case to case. Some enzymes merely provide an environment different from that of the aqueous medium, or they bring the reactants into close contact. Other enzymes take a more active role by adding or subtracting a proton, by straining bonds in the substrate molecule or even by forming transient covalent bonds between the substrate and some part of the enzyme itself. Certain enzymes are helped by the accessory molecules called coenzymes. The coenzyme binds to a specific site on the protein and provides chemical functions that are not available in the enzyme itself.

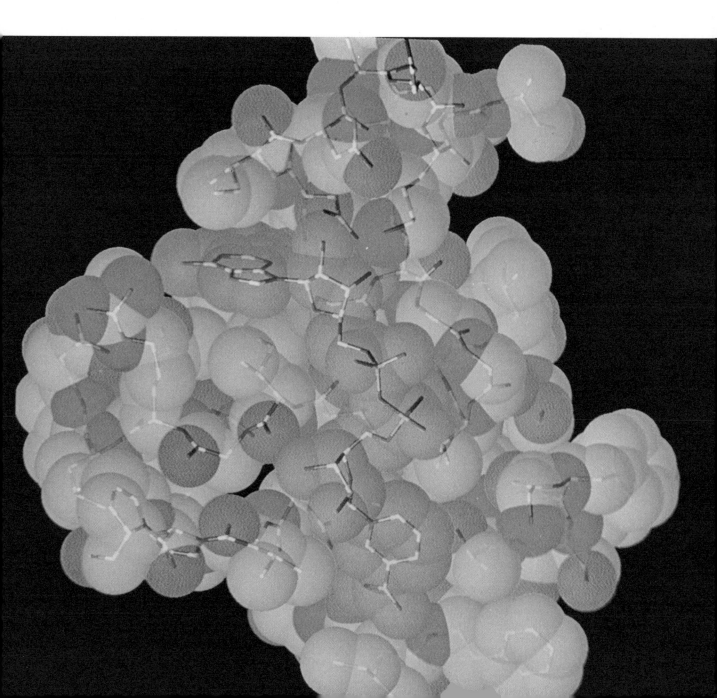

More than 2,000 enzymes have been identified on the basis of the chemical reactions they catalyze. All these proteins must be structurally distinct; in other words, proteins must come in at least 2,000 forms capable of recognizing specific molecules. How are these diverse structures generated? The "alphabet" from which proteins are made consists of the 20 amino acids that can be specified in the genet-ic code; every protein is a sequence of amino acids drawn from this alphabet. The physical and chemical properties of a protein molecule depend on how the chain of amino acids folds up in three-dimensional space.

All the information needed to define the three-dimensional structure of a protein is inherent in the amino acid sequence. As the chain is constructed on the ribosome, it folds up in the way that minimizes the free energy; in other words, the chain assumes its "most comfortable" configuration. In principle, if one knew all the forces acting on the thousands of atoms in the protein and on the surrounding solvent molecules, one could predict the three-dimensional structure from knowledge of the sequence alone. Such a calculation is not now feasible.

The 20 amino acids are all built on a

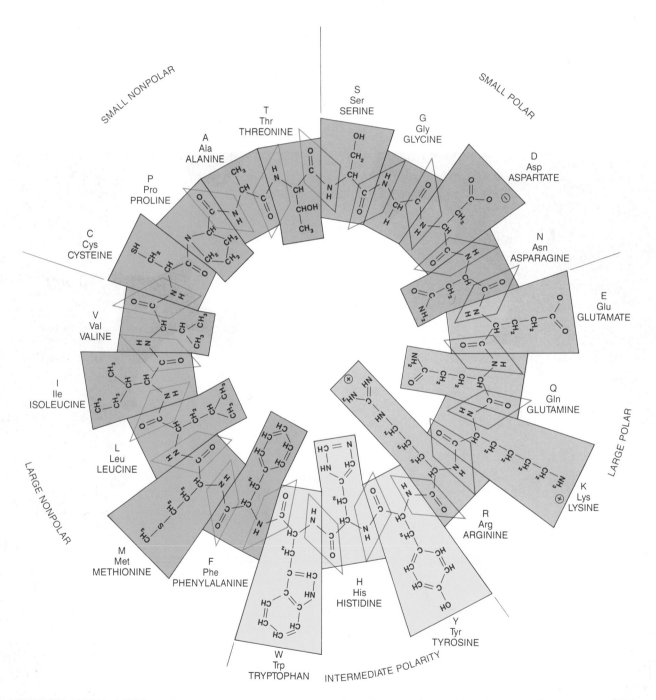

TWENTY AMINO ACIDS specified in the genetic code are the basic components of all proteins. Here the amino acids are shown joined head to tail to form a ring (which is not the structure of any real protein); their three-letter and one-letter abbreviations are indicated. The arrangement places amino acids that have similar chemical properties near one another in the ring. An approximate classification in five groups is based on the size of the amino acid's side chain and on the degree to which it is polarized. (A polar molecule has separated regions of positive and negative electric charge.) These factors have a major influence on the folding of a protein. In the evolution of a protein a mutant form is more likely to be accepted if an amino acid is replaced by one that has similar properties—by one found nearby in the ring. The ring is similar to one proposed by Rosemarie M. Swanson of Texas A&M University.

common foundation. They have an amino group (NH₂) at one end and a carboxylic acid group (COOH) at the other end; both groups are attached to a central carbon atom called the alpha carbon. Also attached to the alpha carbon are a hydrogen atom and a fourth group called the side chain. It is only in the nature of the side chain that the amino acids differ from one another.

The backbone of the protein is built by linking amino acids head to tail: the amino group of one unit is joined to the carboxyl group of the next. The fusion is accomplished by removing a molecule of water, leaving the structure –CO–NH–. The carbon-nitrogen linkage created in this way is called a peptide bond, and the protein chain is referred to as a polypeptide.

The properties of the peptide bond impose certain constraints on the folding of the protein. Electrons are shared among the oxygen, carbon and nitrogen atoms in a way that gives the bond torsional stiffness; it resists rotation about its axis. As a result each peptide-bond unit lies in a plane, and the chain must fold almost entirely through rotations of the alpha-carbon bonds. The polypeptide backbone is not so much a flexible string of beads as it is an articulated chain of flat plates.

The main influence on protein folding comes from the properties of the side chains. Interactions of one side chain with another and with molecules in the medium can force the polypeptide to fold up into a compact globule with a specific, stable shape.

Some of the amino acids are polar molecules: although they are electrically neutral overall, they have localized concentrations of positive and negative charge. The polarization results from the presence of oxygen or nitrogen atoms, which have a strong affinity for electrons. A few of the amino acids not only are polar but also carry a net electric charge; in other words, they are ionized under physiological conditions. Other side chains (generally those made up exclusively of carbon and hydrogen) are nonpolar. There is a strong tendency for the polar side chains to seek a polar environment and for the nonpolar ones to be segregated in nonpolar areas. Water, the medium in which most proteins are immersed, is a strongly polar substance. When a polar or charged side chain projects into the aqueous environment, the water molecules assume an orderly arrangement. A nonpolar side chain in water disrupts this alignment of charges.

The chief consequence of these interactions is that a protein chain tends to fold so that polar side chains are on

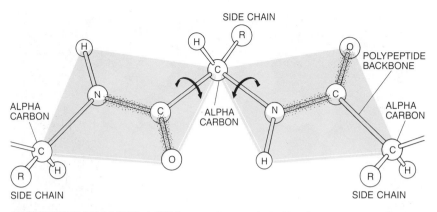

**STRUCTURE OF AMINO ACIDS** constrains protein folding. In an isolated amino acid four chemical groups are attached to the central, or alpha, carbon atom: an amino group (NH₃), a carboxylic acid group (COOH), a hydrogen atom and a side chain (designated *R*). The 20 amino acids differ only in the identity of their side chains. In a protein the amino group of one amino acid is linked to the carboxyl group of another (with the loss of a water molecule), forming a peptide bond. The sharing of electrons among nitrogen, carbon and oxygen atoms makes the bond resistant to twisting, so that each amino acid unit is a rigid plane. The protein can fold only by means of rotations about the bonds to the alpha carbon.

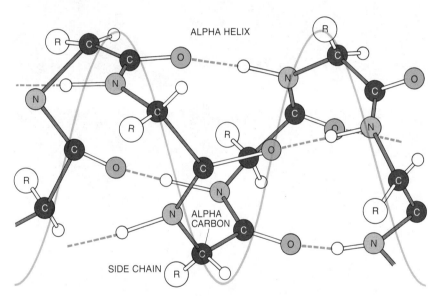

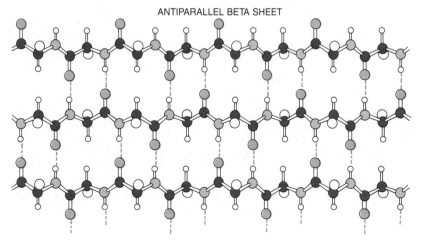

**ALPHA HELIX AND BETA SHEET** are common structural units of protein molecules. The sequence of amino acids in a protein is called its primary structure; as the chain is synthesized, regions of it fold spontaneously into alpha helixes and beta sheets, which constitute the secondary structure; the helixes and sheets are assembled in turn to create the tertiary structure. Both the alpha helix and the beta sheet are stabilized by hydrogen bonds (*broken colored lines*), in which a hydrogen serves as a bridge between oxygen or nitrogen atoms.

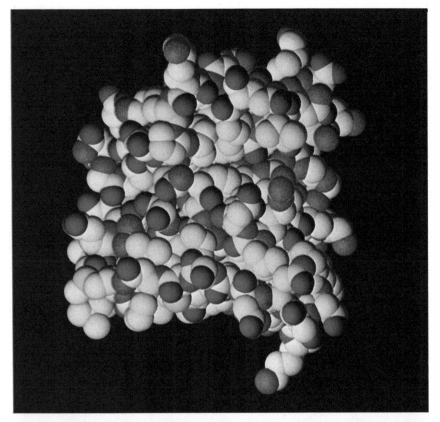

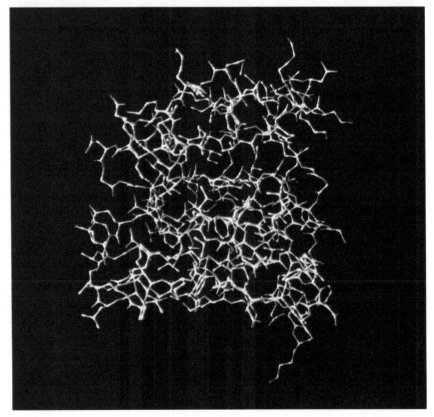

**OVERALL CONFORMATION** of the coenzyme-binding domain of alcohol dehydrogenase is shown in two graphic representations. The space-filling model (*upper image*) emphasizes the surface texture of the molecule; the skeletal diagram (*lower image*) reveals the internal structure. In these views the complete polypeptide backbone and all the amino acid side chains are shown, but hydrogen atoms are omitted. The folding of the chain appears to be a random jumble but is actually quite specific: every molecule of alcohol dehydrogenase folds in exactly the same way. The domain, about half of the molecule, is seen from another point of view in the illustration on page 39. Here the NAD-binding site is at bottom.

the exposed surface and nonpolar ones are inside. An exception to this rule is found in proteins embedded in cell membranes. The membrane is made up of fatty, nonpolar molecules, and the segment of the protein that passes through it likewise consists mainly of nonpolar amino acids. They anchor the protein in the membrane.

The electrostatic attraction between a polar side chain and water is a form of hydrogen bonding, in which a hydrogen atom acts as a bridge between charged oxygen or nitrogen atoms. Hydrogen bonding between one atom and another within the protein itself also helps to stabilize the structure.

Hydrogen bonds are weaker than the covalent bonds of the polypeptide backbone. Moreover, the atoms in a protein that are hydrogen-bonded to one another could as easily be hydrogen-bonded to water; the energy difference between the two configurations is small. Because many hydrogen bonds can form simultaneously as the protein folds, however, they contribute greatly to the stability of the structure.

Still another form of bonding can cross-link regions of the molecule. The amino acid cysteine has a sulfhydryl (SH) group at the end of its side chain. If the protein includes two cysteine units, they can combine to form a covalent disulfide bond (–S–S–). Such cross-links are much stronger than hydrogen bonds.

The amino acid sequence of a protein is called its primary structure. The complete three-dimensional conformation of a single polypeptide strand is referred to as the tertiary structure. As these terms suggest, there is an intermediate level of organization called the secondary structure. It describes the local folding of the chain in terms of structural units that appear in almost all proteins.

Some 35 years ago Linus Pauling showed that the protein backbone can be coiled into a tight helix stabilized by numerous hydrogen bonds; he called the structure the alpha helix. The helix makes one turn for every 3.6 amino acids, and hydrogen bonds form between amino acids four units apart. The bonds do not involve the side chains but rather extend from the NH group of one peptide unit to the CO group of another; for this reason the stability of the helix is not strongly dependent on the identity of the side chains, and many different sequences of amino acids can spontaneously assume the form of an alpha helix.

At about the same time, Pauling proposed a second stable configuration he designated the beta sheet. In this case lengths of polypeptide chain lie next to one another and run either parallel or antiparallel, with hydrogen bonds connecting the adjacent strands. Again the bonds join the NH and CO groups of the backbone.

Some proteins are composed mostly of alpha helix and others are predominantly beta sheet. In a typical globular protein the interior is a bundle of beta strands running back and forth diametrically and the surface is covered with alpha helixes. The exterior helixes generally show a characteristic periodicity in amino acid sequence. Nonpolar side chains appear at every third or fourth position and are directed toward the interior of the molecule; the rest of the side chains, which are exposed to the aqueous environment, tend to be polar.

In recent years still another intermediate level of protein structure has been perceived. For example, a structural element present in numerous proteins consists of two beta strands connected by a segment of alpha helix. The three pieces nestle together comfortably when they are arranged at particular angles. A structural feature of this kind, which typically encompasses from 30 to 150 amino acids, is called a domain. It can be considered a single unit because its conformation is determined almost entirely by its own amino acid sequence. The beta-alpha-beta domain is of particular importance because when two such domains lie next to each other, the crevice they form often serves as a binding site.

A typical globular protein includes about 350 amino acids, which could fold in innumerable ways. The hierarchy of larger-scale structures brings a measure of order. Local interactions between nearby amino acids give rise to alpha helixes, beta sheets or other forms of secondary structure. These subassemblies, acting as more or less coherent units, organize themselves into domains. The geometric arrangement of the domains constitutes the tertiary structure. The presence of the same secondary structures and domains in many dissimilar proteins argues that they are not mere artificial abstractions introduced by the biochemist; on the contrary, they seem to be fundamental units in the evolution and diversification of proteins.

Many proteins have a level of organization beyond the tertiary structure. They are composed of multiple polypeptide strands held together by a variety of weak bonds and sometimes further cemented by disulfide linkages. Some proteins also have nonpeptide components. Metal ions, for example, are essential to the activity of certain enzymes, and a structure called the

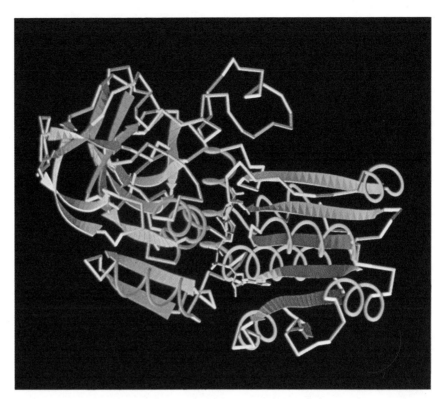

**SECONDARY STRUCTURE** of alcohol dehydrogenase consists of numerous alpha helixes and beta sheets connected by short lengths of "random" structure. The NAD-binding domain is in green and yellow; the catalytic domain, which binds to an alcohol molecule, is in blue. A bound NAD molecule is shown in purple near the junction of the protein domains.

porphyrin ring is found in hemoglobin, chlorophyll and a number of other proteins. Many proteins are also "decorated" on their surface with chains of sugar molecules. These additional features of protein structure are elaborations of the molecule added after the polypeptides are synthesized.

In a way it is remarkable that any protein consistently assumes a single, well-defined conformation. The folded state does have a lower free energy than any alternative configuration, but the difference is small. In an alpha helix hydrogen bonding between peptide units reduces the energy, but if the helix were unraveled, the same sites would form hydrogen bonds with water. Furthermore, because a helix is an ordered structure, it has a low entropy, which tends to increase the free energy. It is worth noting that not all polypeptides have a stable folded pattern. Artificially constructed random sequences of amino acids are generally loose, flexible coils that continually shift from one structure to another. The proteins found in biological systems appear to be a subset of polypeptides selected for their stability of structure.

How is the structure of proteins discovered? Of all the methods employed by the protein chemist, the most re-

vealing has been X-ray crystallography. The basic idea is to form a diffraction pattern by passing X rays through a crystallized specimen of the protein. Because of the periodic structure of the crystal, the pattern is essentially the same as the one that would be generated if a single molecule could be examined. From the diffraction pattern one constructs a map showing the density of electrons in the protein, and from the map the path of the backbone and the positions of the side chains can be inferred.

X-ray crystallography provides a three-dimensional view and shows a protein in atomic detail. It is through such studies that the main themes of protein structure have been elucidated: that the interior is filled with nonpolar side chains, that the alpha helix and beta sheet are more than hypothetical structures, that most proteins are compact globules with dimpled surfaces, and much more. Crystallographers also confirmed the existence of domains and discovered common patterns among them.

Ideally the three-dimensional structure of all proteins would be studied by X-ray crystallography, but that is not feasible. Crystallizing a protein in the first place often calls for a good deal of

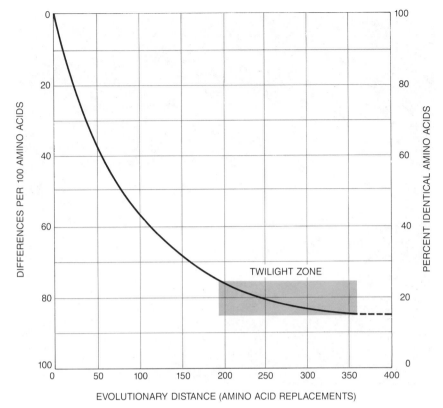

EVOLUTIONARY DISTANCE (AMINO ACID REPLACEMENTS)

**EVOLUTION OF PROTEINS can be traced by comparing amino acid sequences. The number of amino acid positions at which two proteins differ is a measure of their evolutionary distance, but the relation is not a simple one of proportionality. A single site may undergo repeated mutations, so that the actual number of amino acid replacements is generally greater than the number of observed differences. Furthermore, in matching two sequences allowance must be made for insertions and deletions as well as substitutions. As a result, when proteins are identical at fewer than about 15 percent of their positions, common ancestry cannot be distinguished from chance coincidence. Many of the most interesting evolutionary relations lie in the "twilight zone" between 15 and 25 percent identity.**

chemical wizardry, and the subsequent analysis of diffraction patterns is arduous. It took 23 years to map the structure of hemoglobin. Up to now the three-dimensional structures of only about 100 proteins have been solved.

There was a time, in the 1950's, when merely determining the amino acid sequence of a protein was also a difficult and laborious procedure, even for a very small protein. First the total composition of the protein was found by breaking all the peptide bonds in a sample of the material and measuring the amount of each amino acid present. Other samples were only partially digested, leaving small fragments whose amino acid content could then be analyzed in turn. A special chemical trick revealed which amino acid was at the amino end of each fragment. Having gathered information on many overlapping fragments, the biochemist could attempt to solve the elaborate puzzle of how the pieces fit together.

In the 1960's the technology of sequence analysis improved dramatically, and by 1970 the procedure had

been automated. Amino acids were removed one at a time from the amino end of the chain and identified. A limit remained, however, on the maximum length of chain that could be handled, and so large proteins still had to be broken down into fragments.

In the past few years an indirect method of sequence analysis has all but supplanted the traditional techniques of protein chemistry. The key to the new method is that the nucleotide sequence of a DNA molecule is much easier to determine than the amino acid sequence of a protein. If one has a length of DNA that is known to encode the structure of the protein, it is a simple matter to sequence the DNA and translate each three-base codon into the corresponding amino acid.

The one difficult operation is finding the DNA that encodes the protein. In one strategy the first step is to analyze about 25 amino acids at the amino end of the protein. An appropriate segment of from five to seven amino acids within this range is then "back-translated" into a nucleotide sequence. This

process is not without ambiguity: although every codon specifies exactly one amino acid, most of the amino acids can be specified by more than one codon. The key is to choose a sequence that has as little ambiguity as possible and then to produce a DNA molecule for each possible back translation. If the amino acids include a histidine unit, for example, DNA is made with both *CAT* and *CAC* codons at the appropriate position; these are the two codons that specify histidine.

The back-translated DNA serves as a probe to find complementary DNA sequences. It is labeled with a radioactive isotope of phosphorus and allowed to hybridize with the DNA in a library of cloned gene fragments. The clones with a matching sequence are readily identified by the presence of the radioactive phosphorus; they are isolated, cultured in quantity and then sequenced. The approach may seem roundabout, but it is simpler and more accurate than direct chemical analysis of protein fragments. Some 4,000 polypeptide sequences are now known.

The geneticist Theodosius Dobzhansky once wrote: "Nothing in biology makes sense except in the light of evolution." The same is true of protein structure: it makes sense only in terms of protein evolution. Just as all living organisms surely trace their lineage to a few progenitors, the great majority of proteins must be descended from a very small number of archetypes.

Evidence supporting this assertion comes from many quarters, and I shall defend it only briefly. The most straightforward argument is the manifest difficulty of "inventing" a protein de novo. As pointed out above, most random polypeptides do not even fold, much less exhibit a biological function; a new protein is far more likely to arise from modification of an existing one. There is abundant evidence of this process in specific amino acid sequences that are encoded by more than one segment of DNA in a given genome. Moreover, in proteins that fold to form localized domains crystallographers consistently find the same patterns in varied settings; once a substructure has proved useful, it seems to be called on repeatedly.

The primary mechanism of protein evolution is gene duplication, in which a cell comes to include two copies (or more) of a single gene. One copy retains its original function, so that the organism's viability is not compromised by the lack of an essential protein. The redundant copy is therefore free to mutate without constraint from natural selection. Most mutations generate a nonfunctional protein, but an occasional advantageous change can create either an improved version of the original protein or a protein with an entirely new function.

There are two aspects to the study of protein evolution, which must be carefully distinguished. One can examine the "same" protein in various species, observing how the structure has changed over the course of biological time. For example, the amino acid sequence of cytochrome *c,* a protein that transfers electrons in metabolism, has been determined for more than 80 species, from bacteria to man. One product of such studies is a taxonomy of the organisms based on the relations of their proteins. The other approach is to compare the structures of various proteins within a single species. From this endeavor one can construct the family tree of the proteins themselves.

Comparisons from species to species offer considerable insight into protein chemistry. Between closely related organisms the commonest changes substitute one amino acid for another with similar properties, so that the overall structure of the molecule is not disrupted. As the evolutionary distance between the species increases, the sequences diverge. Ultimately the consanguinity of the sequences may be undetectable, even though the two proteins are unmistakably alike in tertiary structure. What this means is that completely different amino acid sequences can fold into the same shape.

In comparing different proteins within a single species it soon becomes obvious there are broad families of related molecules. The half-dozen polypeptides that make up various forms of hemoglobin, for example, and the single polypeptide of myoglobin all share clear similarities. They are not only analogous (meaning they are similar in function) but also homologous (meaning they derive from a common ancestor). Among the enzymes it is not surprising that those catalyzing similar reactions often have homologous sequences. Glutathione reductase and lipoamide reductase provide an illustrative example. Both enzymes catalyze the transfer of hydrogen ions to sulfur-bearing compounds; they are identical at more than 40 percent of their amino acid positions. A similar degree of homology is evident between chymotrypsinogen and trypsinogen and between ornithine transcarbamylase and aspartate transcarbamylase.

As the kinship between proteins grows more remote, sequence homology becomes harder to detect. Worse, the arithmetic of sequence comparison is such that unrelated sequences may appear tantalizingly similar. Offhand, one might expect two randomly chosen polypeptides to be identical at about 5 percent of their amino acid positions; after all, there are 20 amino acids. If the comparison could be made by simply writing down the sequences one above the other and

then ticking off the matches, the 5 percent limit would apply, but in reality a more sophisticated method is needed.

A protein can be altered not only by the substitution of one amino acid for another but also by the deletion or insertion of amino acids. Suppose two proteins are identical except that one has lost its first amino acid; if no allowance were made for this deletion, the proteins would appear to be unrelated. On the other hand, if unlimited gaps and insertions were allowed, any two proteins could be forced to match arbitrarily well. In practice the sequence comparison is done with a computer program that rewards matches between identical or similar amino acids and imposes penalties for gaps and insertions. Even so, it is virtually impossible to distinguish between chance similarity and common ancestry when the number of identical positions falls below about 15 percent.

In tracing the genealogy of proteins the relations of greatest interest are those between sequences that (after adjustment for gaps and insertions) are between 15 and 25 percent identical. This "twilight zone" is where one must look for the roots of the protein family tree, to find molecules that diverged early in the course of their evolution.

In the early 1960's it became clear that a repository of amino acid sequences would facilitate studies of protein evolution, and in 1965 Richard Eck and Margaret O. Dayhoff issued the first volume of the *Atlas for Protein*

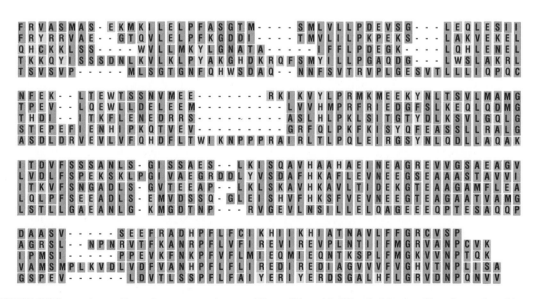

**MOLECULAR PALEONTOLOGY reveals a pattern of common ancestry for five proteins from diverse species. Each protein is represented by a sequence of the one-letter abbreviations for amino acids given in the illustration on page 40; the colors relate amino acids with similar properties. Dashes indicate gaps or insertions. Cysteine units, which can form cross-links that stabilize the folded structure, are marked by boxes. Ovalbumin is an abundant protein** in egg white; antithrombin III and alpha-1 antitrypsin are found in blood plasma; barley protein Z was recently discovered in barley seeds, and angiotensinogen is the precursor of a small protein that regulates blood pressure. Both antithrombin III and alpha-1 antitrypsin are known to act as inhibitors of proteases (enzymes that cut protein chains). The functions of the other proteins had not been known, but now it seems they too may be protease inhibitors.

| BLOOD COAGULATION FACTOR X | ...TREIC - SLDNGG - CDQFCREERSE - VRCSCAHGYVLGDDSKSCVS... |
|---|---|
| EPIDERMAL GROWTH FACTOR PRECURSOR | ...TCTGCSSPDNGG - CSQICLPLRPGSWECDCFPGYDLQSDRKSCAA... |
| LOW-DENSITY LIPOPROTEIN RECEPTOR | ...DIDECQDPD - - - TCSQLCVNLEGG - YKCQCEEGFQLDPHTKACKA... |
| TISSUE PLASMINOGEN ACTIVATOR | ...SCSEP - RCFNGGTCQQALYFSD - - FVCQCPEGFAGKCCEIDTRA... |
| UROKINASE | ...VPSNC - DCLNGGTCVSNKYFSNIHW - - CNCPKKFGGQHCEIDKSK... |
| COMPLEMENT COMPONENT 9 | ...SVRKCHTCQNGGTVILM - - - - - DGKCLCACPFKFEGIACEISKQK... |

**COMMON SEQUENCE** embedded within six disparate proteins suggests they may have shared genetic information at some point in their evolution. Only a segment of each protein is shown; it corresponds to a single identifiable domain. The similarities within the domain are unmistakable, even though some of the proteins differ greatly elsewhere in their structures. It appears that DNA encoding the domain has been copied from gene to gene. The proteins are all recent products of evolution, found only in vertebrate animals.

*Sequence and Structure.* Their goal was to publish annually "all the sequences that could fit between a single pair of covers." It soon became apparent, however, that the covers would have to be very far apart, and computer tapes began to replace bound volumes as the working medium of the sequence comparer. Today any investigator can gain access to large sequence banks from a computer terminal.

About 10 years ago, working with a tape of data from the *Atlas,* I began to study the phylogeny of certain proteins. I was soon maintaining my own data bank, and whenever a new sequence was reported, I would enter it in the archive to see if it resembled anything already known. The number of matches was surprisingly large.

I should like to give an example of how this molecular paleontology works. In the late 1970's Staffan Magnusson and his co-workers at the University of Aarhus in Denmark determined the amino acid sequence of antithrombin III, a protein in the blood plasma of vertebrate animals. Antithrombin III neutralizes thrombin, a blood-clotting factor whose mode of action is that of a protease, or protein-cutting enzyme. At about the same time a second group reported the sequence of alpha-1 antitrypsin, another protease inhibitor in the blood plasma. The Danish group compared the two sequences and found they were identical at 120 of 390 sites, a homology of about 30 percent. It seemed obvious they had descended from a common ancestral protein.

Not long after, workers at the National Biomedical Research Foundation at Georgetown University entered into their computer the sequence of ovalbumin, a protein abundant in egg white. They found that it resembles antithrombin III and alpha-1 antitrypsin, again to the extent of about 30 percent. The discovery came as a surprise, because up to then no one had any idea what the function of ovalbumin might be. The possibility that it is a protease inhibitor now had to be considered.

In 1983 a Japanese group published the sequence of angiotensinogen, the precursor of a small peptide hormone that regulates blood pressure. Although the hormone itself is only 10 amino acids long, the precursor extends to about 400 units. When I compared the sequence of angiotensinogen with the sequences in my data bank, the search revealed a low-level resemblance to alpha-1 antitrypsin. The resemblance was one of those in the twilight zone, amounting to only a 20 percent identity, but a statistical analysis convinced me the two proteins are members of the same family. Since then corroborating observations have been made by others, and there is no doubt of the kinship.

Another Danish group has recently added a fifth branch to this unexpected tree of related proteins: it is a substance of unknown function found in barley seeds and called protein Z. Although protein Z is only half the size of the others (about 200 amino acids), it is clearly related to them. Indeed, the half size fits well with experimental findings that the other proteins in the family have two major domains.

The discovery of these five related proteins in diverse settings suggests two lessons. First, whether or not the 4,000 amino acid sequences known today represent a significant fraction of all proteins, a point has been reached where any newly determined sequence has a good chance of resembling one already on record. Second, certain large-scale arrangements of amino acids are so useful in biochemistry that they have been employed over and over again in different contexts. Often these functional units can be identified with the domains recognized in structural studies.

One of the most widely distributed domains was discovered in 1974 by Michael G. Rossmann and his colleagues at Purdue University. They noted from X-ray-diffraction maps that several enzymes had an important feature in common: even though the overall structures of the proteins were quite different, they all included a domain of about 70 amino acids with essentially the same folding pattern. The enzymes also differed greatly in function, but they had in common the ability to bind certain coenzymes, namely nicotinamide adenine dinucleotide (NAD), flavin mononucleotide (FMN) or adenosine monophosphate (AMP). All these molecules include a mononucleotide within their structure. The ubiquitous domain in the enzymes is the binding site for the mononucleotides, and Rossmann named it the mononucleotide fold.

The discovery led Rossmann to a bold hypothesis. The domain found in all these enzymes, he proposed, is the ghost of a primitive protein from precellular times. Its ability to bind nucleotides was so important that it was incorporated into the machinery of several of the prototype enzymes that emerged in the first living systems. It is still recognizable today.

The model is an attractive one. It seems likely that the first functioning proteins were small and that their important capability was the binding of other molecules. If two small proteins able to bind two different small molecules were joined, the rudiments of catalysis could be initiated. Once started, a succession of gene duplications could lead to an extended family of stable proteins. In the early stages the loosely gathered proteins would be clumsy and inefficient. Opportunities for improvement would be numerous, however, and the natural selection of mutant structures would dominate events. Eventually the proteins would be so artfully suited to their function and the enzymes so efficient that "natural rejection" of mutant variants would prevail.

Not long after Rossmann suggested that primitive proteins might have been created by the fusion of useful domains, recombinant-DNA studies led to the startling discovery that eukaryotic genes are not continuous. They consist of segments that encode part of a protein's structure (exons) separated by long stretches of noncoding DNA (introns). In some cases the introns were observed to fall at or near the boundaries of a protein's domains. This correspondence led Walter Gilbert of Harvard University to propose that exons are the genomic equivalent of the interchangeable protein parts hypothesized by Rossmann. In Gilbert's view, not only were the first proteins created by the assembly of stable domains but also evolution had maintained the genetic isolation of the domains over the course of several billion years. It is easy to see how this genomic organization might convey an adaptive advantage: the continuing reassembly of domains in new combinations would give rise to novel, and occasionally useful, proteins. Gilbert's ideas have been widely accepted, although there are also counterarguments. In many eukaryotic proteins introns fall at places other than obvious domain boundaries. Furthermore, prokaryotic genes have no introns at all; it is necessary to suppose they were eliminated in the interest of genomic economy.

Lately another remarkable instance of the dispersal of domains throughout a group of proteins has come to light. In this case the proteins are all recent products of evolution; they are found only in vertebrate animals that arose well within the past billion years. Moreover, the distribution of the domains among these proteins cannot readily be explained as a simple result of descent from a common ancestor. One domain is present in 18 copies scattered throughout six proteins. It seems clear these subunits have been passed freely from one protein to another and have been inserted wherever their functional activity is needed. For several of the proteins it has been shown that the DNA coding for the domains is precisely delimited by introns. In these cases there can be no question that the organization of the genome into exons and introns has

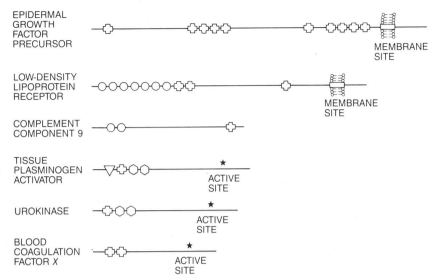

**SHUFFLING OF MULTIPLE DOMAINS is evidence of the continual diffusion of genetic information in higher organisms. Five domains are represented by various geometric symbols, and their distribution is shown in six proteins. The domain whose sequence is given in the illustration on the opposite page is the one marked here by a cross. The distribution of the domains cannot readily be explained by assuming that all the domains in a given protein were inherited from the same ancestral gene; instead they seem to have spread from one protein to another by chromosomal rearrangements. In several cases boundaries between domains in the protein correspond to boundaries between exons and introns in the genome, which may have facilitated the shuffling of gene segments in the course of evolution.**

been instrumental in the rearrangement of the mosaic gene products.

Does this confirm Gilbert's hypothesis that exonic shuffling of domains has been a major feature of protein evolution from the earliest times? Although such shuffling is certainly going on now, I think it is a mistake to assume the same mechanism was at work in more primitive organisms. Introns can be dealt with in eukaryotic genes only because sophisticated splicing machinery ensures that the pieces of messenger RNA are properly translated into

protein. It seems unlikely the same apparatus could have been present in the earliest life forms. Exon exchange in the mosaic vertebrate proteins is more likely a reenactment of ancient events, but in a totally modern guise.

Such variations on a theme are to be expected in a system as complex as the living cell, where a change in one molecule can affect thousands of others, including the very machinery responsible for synthesizing the first molecule. Just as proteins evolve, so do the mechanisms of protein evolution.

# 5

# THE MOLECULES OF THE CELL MEMBRANE

# The Molecules of the Cell Membrane

*They spontaneously form a simple, two-dimensional liquid controlling what enters and leaves the cell. Some cells internalize and then recycle a membrane area equivalent to their entire surface in less than an hour*

by Mark S. Bretscher

The organization of chemical activity in all higher cells depends in large part on the compartmentation afforded by biological membranes. The basic building blocks of membranes are a class of molecules called lipids, which by virtue of their interactions with one another in a watery medium form a closed and flexible compartment. Embedded in the lipid matrix are many different kinds of protein molecules, which give each kind of membrane its distinctive identity and carry out its specialized functions. The primary function of all membranes, then, is to separate what is inside the membrane compartment from the environment outside it. Within the cell, for example, membranes serve to isolate the chemical reactions that take place inside each intracellular organelle. The cell itself is encapsulated by its own cell membrane: the plasma membrane. The plasma membrane is the best-understood membrane, and most of this discussion will be devoted to it.

Evidently if nutrients are to enter the cell or if waste material is to leave it, the materials must somehow cross the barrier created by the lipid matrix of the plasma membrane. The crossing is usually effected by globular protein molecules that span the plasma membrane and catalyze the transfer of specific nutrients and waste molecules. Some of the nutrient molecules required by eukaryotic cells are too large to be transported across the membrane in this way, however. Instead certain protein receptor molecules, anchored by their tail in the plasma membrane, bind these nutrients from the surrounding medium. In a process called endocytosis, pits develop in the membrane and engulf many such receptor molecules and their bound nutrients, which at this stage are called ligands. The pits close up and bud off into the cell, forming vesicles in the cytoplasm, or internal fluid of the cell. At the same time other vesicles from the interior of the cell fuse with the plasma membrane and expel their contents into the surrounding medium. Such pitting and fusing circulates membrane from the surface of the cell to its interior and back again. An area of membrane equivalent to the area of the entire surface of the cell takes part in the cycle every 50 minutes.

The study of the plasma membrane has focused in recent years on the mechanisms that underlie this circulation and on its various effects. Although it has been accepted for some time that the primary function of the endocytic cycle is to bring specific nutrients into the cell, it is now increasingly clear that it can serve the cell in other ways too. For example, my own recent work (which I shall not describe here) suggests that the cell can exploit the endocytic cycle to move about on a substrate.

Another major issue is to understand how each kind of membrane, including the plasma membrane, gains its own unique set of proteins, which determines both its identity and its functions. The problem of membrane identity is complicated by the continual exchange of membrane among the various cellular organelles that takes place, for example, during the endocytic cycle. How, given such mixing of membrane, is the integrity of each set of membrane proteins maintained?

The basic framework of all membranes is a double layer of lipid molecules, an arrangement originally proposed by E. Gorter and F. Grendel of the University of Leiden in 1925. Nature has evolved a variety of lipid molecules all of which share a critical property: one end of the molecule is soluble in water and is chemically described as hydrophilic; the other end is a hydrocarbon, is therefore oily and insoluble in water and is chemically described as hydrophobic.

The commonest membrane lipids belong to a class called the phospholipids. They have a hydrophilic head group made up of a phosphate linked to a residue that can be either choline, ethanolamine, serine or inositol. The head group is attached to two hydrophobic tails, each of which is a fatty acid chain. The most abundant and most widely studied phospholipid is the one having a choline residue. It is called phosphatidylcholine. Like other phospholipids it has a remarkable property: when they are introduced into a watery environment, the individual molecules spontaneously arrange themselves into a bilayer. In the bilay-

BASKETLIKE NETWORK of protein molecules called clathrin coats a closed, spherical piece of membrane called a vesicle, which was isolated from a human placenta. The coated vesicle is derived from the plasma membrane of the cell through a dynamic process called receptor-mediated endocytosis, whereby large molecules are brought into the cell. When selected molecules outside the cell become attached to protein receptors in the membrane, a coat of clathrin begins to assemble itself on the side of the membrane facing the cell interior. Each molecule of clathrin is a chain of about 1,600 amino acids, and the coat is a honeycomb structure formed when the molecules become aligned in a regular pattern. As the coat grows, the region of membrane to which it is attached bulges into the cell in a way that resembles the formation of a drop of water on the lip of a faucet. The bulge pinches off from the surface of the cell and becomes a vesicle whose inside surface carries the receptors and their ligands and whose outside surface retains the honeycombed coat of clathrin seen in the image. The image was constructed by computer from a series of electron micrographs made at various tilt angles by Guy Vigers of the Medical Research Council's Laboratory of Molecular Biology in Cambridge. Enlargement is more than two million diameters.

er the molecules in both layers align themselves in such a way that their longest axis is roughly perpendicular to the plane of the bilayer. The hydrophilic head groups face water on both sides of the bilayer, and the oily, hydrophobic tails sequester themselves in the middle of the bilayer, thereby excluding water from it. The arrangement is the state of lowest free energy for these molecules in water.

In 1965 Alec D. Bangham and his colleagues at the Agricultural Research Council's Institute of Animal Physiology in Cambridge showed that phospholipid bilayers in water form closed spherical vesicles having two separated compartments: the fluid in-side the vesicle and the fluid outside. Such vesicles form because if a free edge on a bilayer were exposed, some of the hydrophobic regions of the phospholipid molecules would be in contact with water; that would be energetically unfavorable. It is this property of lipids that makes them so effective in biological systems: they spontaneously form a closed envelope with considerable mechanical strength.

Two general features of the bilayer are important in the formation of a biological membrane. First, because they have a hydrocarbon interior, they are essentially impermeable to most biological molecules, such as amino acids, sugars, proteins and nucleic ac-ids, and to ions. All are highly soluble in water and insoluble in hydrocarbon solvents. It is this feature that enables the bilayer to function as a barrier.

Second, a bilayer formed from naturally existing phospholipids is a liquid. There is a double sense in which the bilayer exhibits the random motions characteristic of the liquid phase. The hydrocarbon tails of the phospholipid molecules wiggle about, and so the bilayer is soft and flexible—with the viscosity of, say, olive oil rather than paraffin wax. Furthermore, the molecules can diffuse sideways freely within their own monolayer, and so two neighboring phospholipids in the same monolayer can change places with each

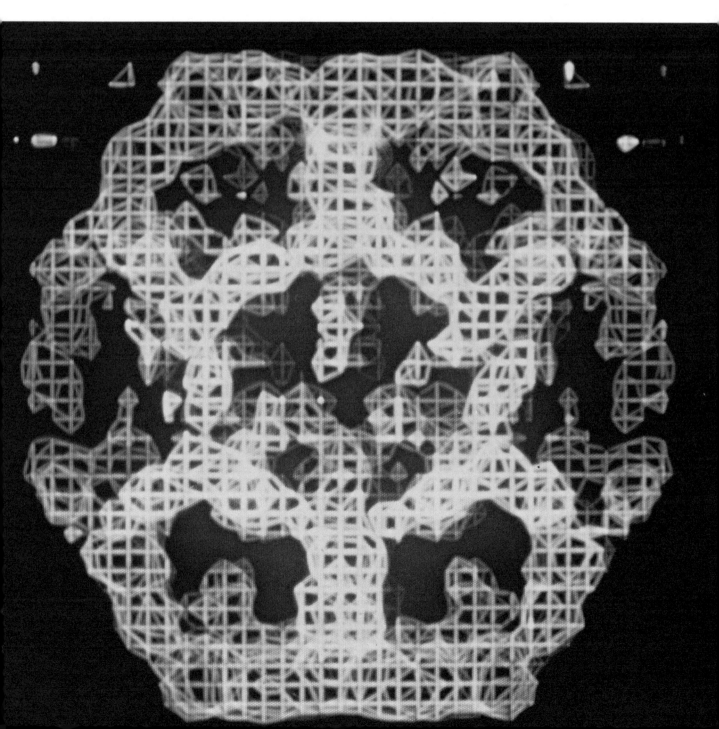

other about once every microsecond. The phospholipid molecules in opposite monolayers, however, almost never change places: such an exchange is made, on the average, only about once a year. Hence each monolayer is a two-dimensional liquid. Physiologically the liquid nature of bilayers is quite important. If the bilayer were a rigid structure, for example, the nerve cells in the neck would crack whenever a person nodded.

In a natural membrane one might expect to find the various kinds of phospholipid molecules randomly distributed on both sides of the bilayer, but in 1972 I discovered that the distribution is much more orderly. In the plasma membrane of the red blood cell I found the outer monolayer includes only phosphatidylcholine and its close relative, sphingomyelin, both of which contain choline. In contrast, the monolayer facing the cytoplasm has phosphatidylethanolamine and phosphatidylserine. It is thought that phosphatidylinositol also resides on the cytoplasmic side of the bilayer.

In addition to the phospholipids two other kinds of lipids are found in the membranes of animal cells: glycolipids and cholesterol. The glycolipid molecule has a hydrophobic tail similar to that of sphingomyelin. As its prefix implies (glyco- is from the Greek word for sweet), the glycolipid's hydrophilic end is composed of a variety of simple sugars joined to form a linear or branching structure called an oligosaccharide. Glycolipids make up only a small fraction of the lipids in the membrane, and they are confined to the outer monolayer.

Cholesterol, on the other hand, is (together with phospholipid) a major membrane lipid. It is a large, disk-shaped molecule with four carbon rings that are fused together, giving the molecule a rigid structure. One end of cholesterol is hydrophilic, but the rest of it is hydrophobic and embeds itself in the hydrophobic part of the plasma membrane. Roughly equal numbers of cholesterol and phospholipid molecules are in the plasma membrane of eukaryotic cells. The addition of cholesterol to the phospholipid matrix makes the membrane somewhat less flexible and even less permeable.

Several puzzles about the lipids in the plasma membrane are still unsolved. The biological role of the glycolipids, for example, is not yet known. Nor is there yet any convincing explanation for the distribution of phospholipids in the bilayers. Why are the bilayers of a eukaryotic cell made up of a variety of phospholipids rather than of only, say, phosphatidylcholine? What is the function of the phospholipids' asymmetric distribution? Finally, the geometry of the bilayer itself presents a problem. The two monolayers are essentially independent of each other, but of course they cover the same area. What are the lateral forces in each monolayer, then? Is one monolayer under compression and the other under tension, or is the lateral pressure the same in both?

Whereas the lipids form the matrix of a membrane, the proteins carry out all its specific functions. The membrane proteins can be classified roughly into two general kinds according to their shape within the hydrocarbon core of the membrane. The shape of one kind is a rodlike, tightly coiled spiral called an alpha helix. In this structure the amino acids that make up the polypeptide chain are so arranged that the protein backbone is a helix and the amino acid side chains project outward from the helix. The second kind of membrane protein appears to have a substantial globular structure within the membrane's hydrophobic region.

One of the clearest examples of a membrane protein with the alpha-he-

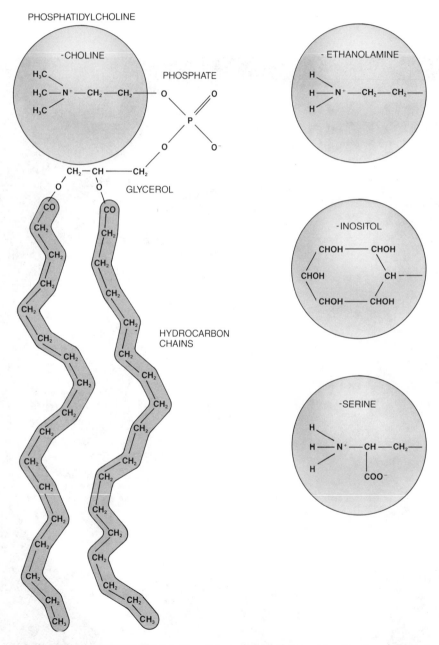

**PHOSPHOLIPID MOLECULE** is the primary structural element in all cell membranes. Four main kinds of phospholipid are found in animal-cell membranes. The one shown at the left in the diagram is phosphatidylcholine, but the other three differ from it and from one another only in the chemical structure of their head groups, which are diagrammed here as colored spheres. The electric charge in each head group makes the group hydrophilic. The head group is connected to a glycerol group, and two hydrocarbon chains are attached in turn to the glycerol. The hydrocarbon chains are oily and therefore hydrophobic.

lical structure is glycophorin, the major glycoprotein of the red blood cell. Although its function remains enigmatic, its structure is now quite well known. Most of the molecule resides on the outside of the cell. This extracellular region is a long sequence of amino acids to which hydrophilic oligosaccharide chains are attached.

In 1971 I showed that glycophorin spans the cell membrane. Two years later Vincent T. Marchesi, who was then at the National Institute of Arthritis, Metabolism, and Digestive Diseases, suggested the geometry of its intramembrane domain. He and his colleagues determined the amino acid sequence of the protein and found that its extracellular region is attached to a segment of 26 hydrophobic amino acids. The 26 amino acids are joined in turn to a short hydrophilic tail. The hydrophobic sequence is just the right length to span the bilayer as an alpha helix, and the short hydrophilic tail rests in the cytoplasm to anchor the protein in the bilayer.

Many other kinds of membrane protein are now known to be fixed to the cell surface by a single hydrophobic alpha helix and anchored in the cytoplasm by a hydrophilic tail. Typically they function as receptors for extracellular molecules or as highly specific markings (such as the major transplantation antigens H2 in mice and HLA in humans) that enable the immune system to distinguish foreign invaders from cells belonging to the organism. Other proteins in this class include the surface immunoglobulin receptors on B lymphocytes and the spike proteins of many membrane viruses. Since the functioning of such proteins depends primarily on their extracellular domain, the intramembrane structure need not be extensive.

Perhaps not surprisingly, the globular structure of the second kind of membrane protein is associated with functions requiring a substantial structure within the plane of the lipid bilayer. For example, one of the most abundant proteins in the membrane of the red blood cell is a globular transport protein called the anion channel. As its name implies, the protein catalyzes the passive exchange of negatively charged ions such as chloride or bicarbonate between the blood plasma and the cytoplasm of the cell. How does such a protein function? One early scheme suggested the protein might bind the ion or molecule to be transported on one side of the membrane, diffuse across the membrane and release it on the other side. Another scheme proposed that the protein molecule might rotate within the mem-

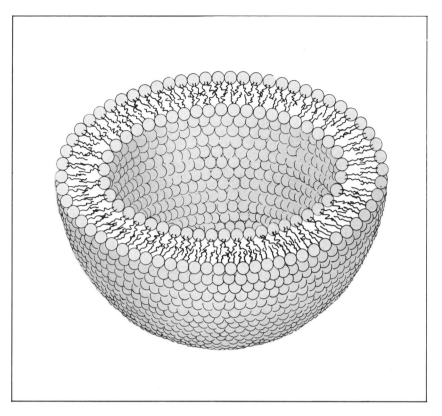

**MOLECULAR ARCHITECTURE** of the animal-cell membrane is determined primarily by the interactions of phospholipid molecules in water. Phospholipids can minimize their energy in water by forming a bilayer about 40 angstrom units thick. The hydrophobic tails of the molecules sequester themselves on the inside of the bilayer and the hydrophilic heads (*blue*) face the water on both sides of the bilayer. If any edge of the bilayer were open to the water, hydrophobic tails along the edge would be exposed; hence the bilayer closes to form a vesicle, effectively segregating fluid inside the vesicle from fluid surrounding it.

brane, thereby bringing the binding site and its attached substrate from one side of the membrane to the other.

Neither view turned out to be correct. In 1971 I showed that what is now known to be the anion channel spans the membrane bilayer and has a fixed and unique orientation in it. It is now thought there is a small passageway for anions through the protein, which enables them to cross the bilayer.

One of the best-understood globular membrane proteins is bacteriorhodopsin, which straddles the membrane of the bacterium *Halobacterium halobium*. The halobacterium, or salt-loving bacterium, lives in the salt beds of San Francisco Bay. The bacteriorhodopsin in the bacterial membrane is a proton pump: it captures photons from sunlight and exploits their energy to pump protons across the membrane against an energy gradient. The proton gradient generated by the pumping represents potential energy, which later serves to drive the synthesis of adenosine triphosphate (ATP). The breakdown of ATP provides energy for the bacterium's biosynthetic pathways.

The structure of bacteriorhodopsin was determined in 1975 by Nigel Un-

win and Richard Henderson, who were then at the Medical Research Council's Laboratory of Molecular Biology in Cambridge. Their model shows that the polypeptide chain zigzags seven times across the bilayer. Each transmembrane segment is an alpha helix and the helixes are packed together to form a globular structure. The photon is captured by a molecule called retinal (a relative of vitamin A), which is attached to the protein by a covalent bond. The mechanism whereby the energy of the photon is directed to the transport of protons is still not known.

In eukaryotic cells it is a general rule that all membrane proteins carry an oligosaccharide chain (or several chains) on their extracellular domains, just as glycophorin does. The function of the oligosaccharide chains is obscure, as it is in the case of the glycolipids. In addition all membrane proteins, both globular and alpha-helical, are held in place in the bilayer by the same kinds of forces that hold the lipid molecules there: the amino acid side chains of the protein in contact with the hydrophobic lipid chains are also hydrophobic, whereas the other parts of such proteins are hydrophilic. The hydrophilic parts are exposed to water on each side of the bilayer.

Because all membrane proteins reside in a liquid bilayer, they can diffuse sideways just as the lipid molecules do. How fast they diffuse is determined in part by how liquid the phospholipid matrix is. In 1974 Mu-ming Poo and Richard A. Cone, then at Harvard University, showed that rhodopsin diffuses about 10 micrometers in one minute. It seems likely that most other membrane proteins diffuse at about the same rate. Unless they are constrained from doing so, membrane proteins in most eukaryotic cells can therefore diffuse, on the average, from one end of the cell to the other in a few minutes.

The constraints on the diffusion of molecules across the membrane are probably best exemplified in cells that are joined to one another to form an epithelial sheet. They include the cells that line the gut, the dividing cells of the skin and the cells of internal organs such as the liver, kidney and pancreas. Epithelial sheets are only one cell thick; often they are folded extensively to form a compact organ.

Epithelial sheets have two surfaces. In the gut, for example, one surface (the apical surface) faces the digestive tract and the other (the basolateral surface) faces the blood. Because the epithelial gut cells must transport useful materials—and only useful materials—from the intestine to the blood, the cells making up the epithelial sheet must be held together tightly, with no spaces between them. The cells are therefore joined by a set of so-called tight junctions.

The tight junction can be pictured as a circular belt or gasket that lies in the plasma membrane. The belt not only prevents leaks (even leaks of ions) but also separates the cell's plasma membrane into two domains: the apical surface and the basolateral surface. Membrane proteins can wander at random within their own domain, but the tight junction keeps them from moving from one domain to the other.

The separation between the two parts of the membrane maintains the functional asymmetry needed to transport material in only one direction. For example, on the apical surface of the epithelial sheet in the gut each cell carries proteins that channel sodium from the gut into the cell. On the basolateral membrane there is a different set of proteins that pump the sodium out of the cell into the blood. The net result is an extremely selective transfer of sodium ions across the epithelial sheet, and it is accomplished because the specific proteins needed for each step in the transfer are concentrated on the part of the membrane sur-

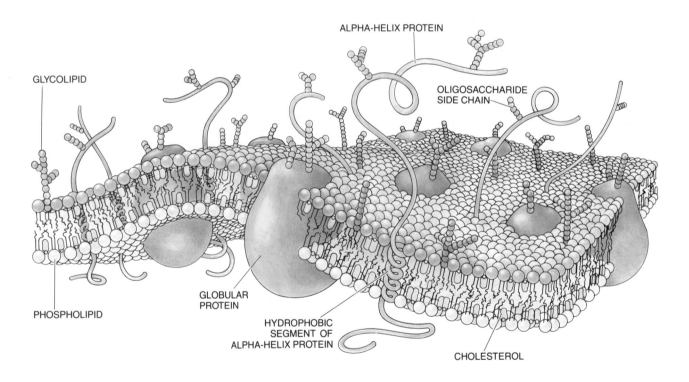

ALPHA-HELIX PROTEIN

GLYCOLIPID

OLIGOSACCHARIDE SIDE CHAIN

PHOSPHOLIPID

GLOBULAR PROTEIN

HYDROPHOBIC SEGMENT OF ALPHA-HELIX PROTEIN

CHOLESTEROL

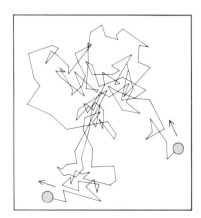

PLASMA MEMBRANE is a phospholipid bilayer in which cholesterol and various kinds of protein molecules are embedded. In this schematic diagram of the membrane the phospholipid molecules in the top layer, which faces the external medium, are shown as dark blue spheres each having two wiggly tails. The chaotic Brownian motion of the molecules within the monolayers is indicated by the diagram at the left; the fluidity of the hydrocarbon interior is suggested by the random configurations of the tails. The bottom layer, which faces the cytoplasm inside the cell, has a different phospholipid composition and is shown in light blue. Although a random exchange of phospholipid molecules also takes place across the bilayer, the event is extremely rare. Two main kinds of protein in the membrane traverse the bilayer. One kind makes the crossing as a single chain of amino acids that is coiled into a so-called alpha helix (*orange*); the intramembrane portion of the second kind of protein is globular in structure (*red*). For clarity the ratio of phospholipid to protein is much larger here than it is in a natural membrane. Rigid cholesterol molecules (*yellow*) tend to keep the tails of the phospholipids relatively fixed and orderly in the regions closest to the hydrophilic heads; the parts of the tails closer to the core of the membrane move about freely. Side chains of sugar molecules attached to proteins and lipids are green.

face where they can properly carry out their role.

How are tight junctions formed, and how do epithelial cells sort their membrane proteins into two domains? The first question is still unanswered, but the second is beginning to yield to attacks based on a discovery by Enrique Rodriguez Boulan and David D. Sabatini of the New York University School of Medicine in 1978. They found that when an epithelial sheet growing in culture is infected with influenza virus, the progeny viruses emerge only from the apical surface of the sheet. On the other hand, a virus called vesicular stomatitis virus (VSV), which causes a mild disease in cattle, emerges only from the basolateral surface. In order to leave its host cell a virus must assemble a protective coat, and so Rodriguez Boulan and Sabatini concluded that the cell directs the coat proteins of the influenza virus to the apical surface and the coat proteins of VSV to the basolateral surface. The two viruses therefore constitute an experimental system in which the development of asymmetry in epithelial cells can readily be studied.

The cells that make up an epithelial sheet can also be joined to one another by a so-called gap junction. The gap junction is rather like two studs pressed together with a hole through their middles [see bottom illustration on next page]. The hole allows neighboring cells to communicate and coordinate their activities. Small molecules whose diameter is less than about 20 angstroms can pass freely from the cytoplasm of one cell through the pipe formed by the gap junction and into the cytoplasm of an adjoining cell.

The structure of gap junctions has been elucidated by Unwin, now at Stanford University, and his colleagues. Their work shows that each junction is made up of 12 protein subunits, six from each cell. Each group of six is arranged in a hexagon in the plasma membrane of each apposed cell; the two hexagons lock into each other to form a channel between the cells. The channel can be held open or closed, but precisely how such control is achieved is not known. Gap junctions often interact with one another to form a raft, or a large group of junctions, on the cell surface. The aggregate size of the rafts and their confinement to the membrane regions between two cells make it likely that gap junctions are relatively motionless within the liquid bilayer.

Until this point I have carefully avoided any detailed discussion of the many membranes in addition to the plasma membrane that are found in

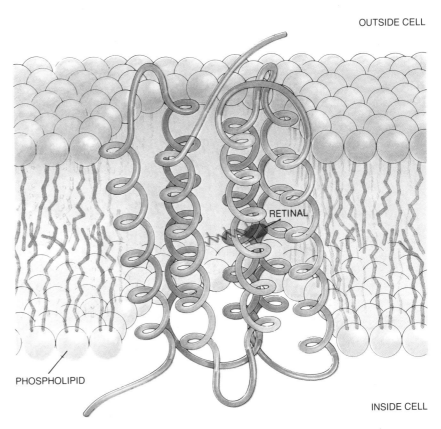

GLOBULAR MEMBRANE PROTEIN bacteriorhodopsin is made up of seven largely hydrophobic sequences of amino acids, joined by short hydrophilic ones. On the basis of findings by Nigel Unwin and Richard Henderson, who were at the Laboratory of Molecular Biology when the work was done, it is now thought that each of the seven hydrophobic sequences is an alpha helix (red) embedded in the hydrocarbon core of the membrane and that the hydrophilic sequences link the helixes to one another on each side of the membrane (blue). A molecule called retinal (green) is attached to the middle of one helix. Retinal captures solar photons, triggering the protein to pump protons across the membrane of certain salt-loving bacteria. The process sets in motion an unusual kind of photosynthesis. The blue spheres and their tails represent the phospholipid bilayer of the bacterial membrane.

the eukaryotic cell. Recall, nonetheless, that many intracellular organelles are defined by a limiting membrane and that such membranes play an essential role in the transport, communication and orderly processing of chemical substances and information within the cell.

Some of the main intracellular organelles take part in the manufacture of membrane components. Membranes are assembled in the endoplasmic reticulum, and oligosaccharides are added to membrane proteins in the Golgi apparatus. Hence the relations among many of the organelles are far from static. For example, there is a continual transfer of membrane from the endoplasmic reticulum to the Golgi apparatus and from there to the plasma membrane. The transfers are probably always mediated by phospholipid vesicles.

Such continual movements of membrane material, as well as the mergers and dissociations of vesicular membranes that accompany the movement, raise anew the question of membrane

integrity. How can specific membrane proteins, destined for or belonging to a specific organelle, avoid mixing and homogenizing during a transfer? A general and precise answer to the question is not yet available, but there is no doubt that what is transferred is not a random sample of the donor membrane. There is one process of material transfer involving two membranes for which the way this is accomplished is beginning to come into focus. That process is endocytosis.

Animal cells obtain most of the small molecules they need for growth either by synthesizing them or by importing them from the blood. The imported molecules are usually transferred across the plasma membrane by specific protein channels or pumps. There are some essential nutrients, however, that for one reason or another cannot be so easily absorbed.

For example, cholesterol (which is needed for the synthesis of membranes) and the ferric ion (an iron atom carrying three positive charges, which is needed for the synthesis of the large, pigmented molecules called cytochromes) both circulate in the blood as large complexes. Cholesterol circulates in the form of cholesteryl esters, which make up the hydrophobic core of a particle called low-density lipo-protein (LDL) that is some 200 angstroms across. Ferric ions in blood are bound inside a large carrier protein called transferrin. Both LDL and transferrin are much too large to pass through a small channel or pump, and so the cell must adopt a radically different strategy to obtain the nutrients it requires.

The current picture of how such nutrients enter the cell began to emerge in 1964 with the work of Thomas F. Roth and Keith R. Porter, who were then at Harvard. They were studying how growing oocytes (egg cells) of the mosquito build up the oocyte's yolk. Examining thin sections of oocytes under the electron microscope, they found that the yolk precursor is bound to the plasma membrane of the oocyte at sites where the membrane is indented and appears to have a thick, dark coat of material on the side facing the cytoplasm. The sites are called coated pits. In the same thin sections of the oocyte Roth and Porter also saw vesicles inside the cell that were full of yolk precursor and had thick coats on their outer surface. They called these structures coated vesicles. The coated vesicles arise when coated pits bud into the cell; they are intermediates, inside the oocyte's cytoplasm, in the transfer of yolk precursor from the cell exterior to the large yolk granules stored inside the oocyte.

More recent work by Richard G. W. Anderson, Michael S. Brown and Joseph L. Goldstein of the University of Texas Health Science Center at Dallas on the uptake of LDL and by many other groups, including my own, on the uptake of transferrin and other large molecules has by now drawn a fairly coherent picture of the early stages of endocytosis initiated by coated pits. On the outer surface of most growing animal cells there are specific protein receptors for LDL, for transferrin and for other large imported molecules. As the receptors diffuse across the surface of the cell they can bind LDL or transferrin.

When an LDL or transferrin receptor encounters a coated pit, it enters the pit. Other proteins in the plasma membrane, however, are excluded from the pit, which thereby acts as a molecular sorting device. In about a minute the coated pit has reached its full diameter of about .3 micrometer. The pit then invaginates and breaks away from the plasma membrane into the cytoplasm, where it forms a coated vesicle. The mechanical force driving the process is assumed to be provided by the coat on the cytoplasmic side of the pit.

Once the coated vesicle has formed in the cytoplasm it sheds its coat in a few seconds. Then two things happen. The vesicle fuses with an intracellular organelle called an endosome and the acidity inside the endosome is then increased to a *p*H of about 5. The acidic environment causes LDL to fall

**EPITHELIAL CELL** and the adjacent parts of its two nearest neighbors are depicted schematically. Such cells line the gut and form dividing layers in other internal organs; in the gut they form a leakproof barrier between the gut and the blood. The seal is effected by the tight junction, which also separates the apical surface facing the gut from the basolateral surface facing the blood. Below the tight junction is a desmosome, which welds the two adjacent cells together, and below the desmosome is a gap junction, which allows small molecules to pass from the cytoplasm of one cell directly into the cytoplasm of the adjacent cell. Nutrient material in the gut can cross the epithelial sheet only if it is first absorbed into an epithelial cell. Some macromolecules can be taken up by endocytosis. The endocytic vesicle can then discharge its contents into the bloodstream by exocytosis on the cell's basolateral surface. The apical membrane and basolateral membrane have different sets of proteins.

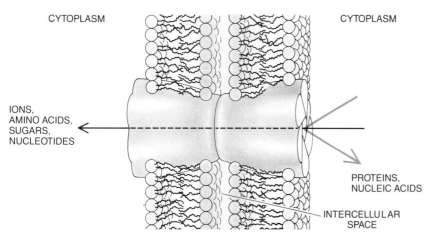

**GAP JUNCTION** between two apposed epithelial cells is made up of two hexagonal studs (*gray*), each embedded in the membrane bilayer of one cell (*blue spheres with tails attached*); the two studs are pressed together in the gap between the cells. Ions, amino acids, sugars, nucleotides and other molecules smaller than about 20 angstroms in diameter can pass through the junction, but proteins, nucleic acids and other larger molecules cannot.

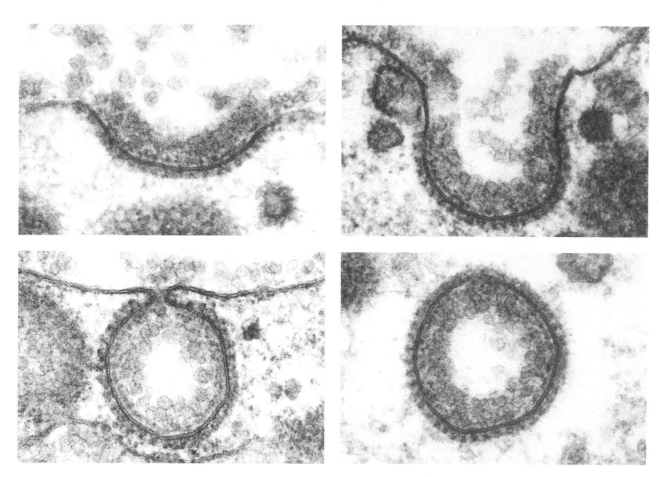

**SUCCESSIVE STAGES** in the formation of a coated vesicle are shown in a series of electron micrographs. The shallow indentation in the plasma membrane of a developing chicken oocyte (*top left*) is a coated pit; it holds many particles of a lipoprotein gathered from the external environment of the cell. A coat of clathrin molecules can be seen just under the pit, on the cytoplasmic side of the membrane. The pit deepens (*top right*), the outer membrane of the cell closes behind the pit (*bottom left*) and the pit buds off to form a coated vesicle that carries the lipoprotein molecules into the cell (*bottom right*). The micrographs were made by M. M. Perry and A. B. Gilbert of the Agricultural Research Council's Poultry Research Centre in Edinburgh. The enlargement is 135,000 diameters.

off its receptor and the ferric ions to pop out of transferrin. By unknown processes the receptors for LDL and for transferrin (the latter with its ligand, transferrin, still attached) are recycled to the plasma membrane. At the same time the LDL, the ferric ions and other contents of the endosome are transferred to lysosomes, again by vesicular transport. The lysosome is a primitive digestive organelle, and it degrades the LDL, thereby liberating cholesterol to serve the needs of the cell. Note that at this stage both the cholesterol and the ferric ions must still be transported across at least one membrane, namely the lysosomal membrane, in order to reach their destinations within the cell.

The endocytic cycle initiated by coated pits gives a dynamic picture of a cell. At any instant about 2 percent of the surface of a cell growing in culture is taken up by deepening coated pits. Given such a large flux of membrane from the plasma membrane through the endosomal compartment and back again, one might expect that the pro-

tein components of the two membranes would quickly become identical. Such mixing does not take place, however, because the coated pits select only certain proteins from the plasma membrane for transfer into the cell. It is thought the same set of membrane proteins, now residing in the endosomal membrane, is cycled back to the plasma membrane by a similar selective process. This selective transfer of membrane by coated pits may explain how the integrity of numerous distinct membrane compartments can be maintained in spite of continual traffic among them. Note also that during the endocytic cycle the topology and asymmetry of the membrane are always maintained.

The understanding of how the coated pit selects proteins from the plasma membrane is admittedly incomplete. Nevertheless, it has been greatly advanced by structural studies of the closely related coated vesicles. In 1976 Barbara M. F. Pearse of the Laboratory of Molecular Biology isolated coated vesicles and showed that the coat is

a lattice of a large, fibrous protein she named clathrin. Such coated vesicles also carry receptors, the ligand molecules attached to the receptors and a variety of other proteins that could mediate the interaction of clathrin with the receptors. Coated vesicles are generated not only by the plasma membrane but also by intracellular organelles such as the Golgi apparatus. Hence there is hope that the sorting mechanism may soon be clarified.

The picture of the plasma membrane emerging from this work is that of a lipid bilayer spanned by a host of different proteins. Some of them simply catalyze the transfer of small molecules across the bilayer, and they have a globular structure. Others have only a single hydrophobic helical segment that holds them in the membrane; some of the helical proteins are receptors that bring large molecules into the cell. Whereas all, or almost all, of these molecules are free to diffuse around in the liquid bilayer, there are other structures such as the gap junctions and the tight junctions that remain relatively static. In contrast there is also the highly dynamic movement of the membrane during the endocytic cycle.

Plasma membranes take part in many cellular functions I have not discussed. Since they make up the interface between a cell and the rest of the organism, they must be involved in the movement of cells and in how the movement is directed during growth and development. The plasma membrane also plays a role in cancerous growth, in which cell multiplication and migration can become uncontrolled. Although a molecular understanding of such processes is yet to be achieved, current knowledge of the structure of membranes is a major step in that direction.

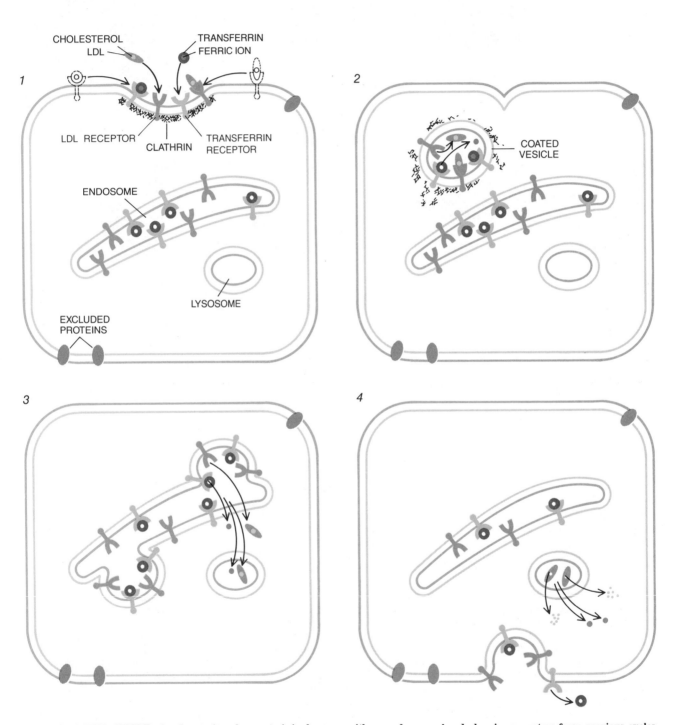

**GENERALIZED CYCLE** of endocytosis and exocytosis is shown in four stages. Ferric ions in transferrin molecules and cholesterol in particles of low-density lipoprotein (LDL) bind to receptors in the plasma membrane. The receptors bearing transferrin and LDL diffuse into a coated pit, which somehow blocks the entry of other kinds of membrane proteins (1). After a pit buds into the cell and becomes a coated vesicle the clathrin coat is shed, and increasingly acidic conditions begin to release LDL from its receptor and the ferric ions from the transferrin (2). The liberated vesicle then fuses with an endosome already bearing receptors from previous cycles of endocytosis. The released ferric ions and LDL are transferred to a lysosome. A vesicle is also shown budding away from the endosome (3). The vesicle, bearing empty LDL receptors and iron-free transferrin still attached to its receptor, then fuses again with the plasma membrane, and the receptors enter another cycle of endocytosis. In the lysosome cholesterol is released from the LDL; ferric ions and cholesterol are transported to other parts of the cell (4). The asymmetry of the membrane is maintained throughout the cycle.

# 6

# THE MOLECULES OF THE CELL MATRIX

# The Molecules of the Cell Matrix

*Proteins in the cytoplasm form a highly structured yet changeable matrix affecting cell shape, division and motion, and the transport of vesicles and organelles. It may also have a bearing on metabolism*

by Klaus Weber and Mary Osborn

The cytoplasm of a cell is not a formless mass of jelly in which the nucleus and the other organelles are scattered. It is highly structured. A fibrous matrix of proteins spans the cytoplasm between the nucleus and the inner surface of the plasma membrane, helping to establish the cell's shape and playing a role in cell locomotion and division. Known as the cytoskeleton, the matrix may also influence the movement of organelles within the cell and even affect cell metabolism by providing a three-dimensional setting for the molecular events that are the cell's vital processes.

Microscopists first observed fibers in the cytoplasm of a few cell types during the 19th century; later silver and other stains were used to enhance the visibility of cytoskeletal fibers under the light microscope. In the 1950's and 1960's electron microscopy revealed three distinct systems of filaments in the cytoplasm. More recent biochemical and immunological studies have identified the distinctive set of proteins characterizing each filament system.

Both ultrastructural appearance and biochemistry distinguish the three systems. The finest of the fibers are the microfilaments; they average six nanometers (billionths of a meter) in diameter and are made up of the protein actin. The microtubules are 22 nanometers in diameter and consist of tubulin. The diameter of intermediate filaments, the third of the systems, ranges between seven and 11 nanometers; their constituent protein varies according to cell type.

The proteins that characterize each filament system can be purified in the laboratory, making it possible to use the technique of immunofluorescence microscopy to gain an overview of the configuration of each filament type within the cell. First applied to a component of the cytoskeleton (the microfilament system) in 1974, the technique requires an antibody to a specific protein. The antibody is made by injecting the purified protein into an experimental animal. One then allows the antibody to bind to the target protein in a fixed cell by introducing the antibody through holes made in the plasma membrane. Next one exposes the cell to a second antibody that recognizes the first and that has been labeled with a fluorescent compound. When the cell is viewed under the fluorescence microscope, the bound fluorescent antibody highlights the distribution of the first antibody and thereby of the protein to which it is bound.

When the first antibody is specific for actin, the fluorescent marker reveals an array of "stress fibers": long, thick bundles of microfilaments running parallel to and just inside the plasma membrane. Stress fibers are particularly prominent in cells that are well spread: cells that lie flat on the substrate. Many normal cells and cells that have been transformed by a tumor virus are rounded in shape and lack stress fibers. The antibody also reveals a fine mesh of microfilaments under the membrane; the mesh is most distinct at the fixed cell's "ruffling edge," which marks the leading edge of a living cell crawling across a substrate.

An antibody that is specific for tubulin makes visible a very different organization. The microtubules it highlights are not concentrated near the plasma membrane; instead they radiate from an organizing center near the nucleus known as the centrosome or cytocenter. The centrosome consists of two compact, cylindrical structures known as centrioles and the surrounding material. Individual microtubules run from the centrioles to just under the plasma membrane. In contrast to the stress fibers the radial microtubules do not disappear when a well-spread cell assumes a more rounded shape, although they must change in length. Under the fluorescence microscope it is more difficult, however, to resolve neighboring microtubules in a rounded cell than it is in a well-spread one, which may account for the erroneous belief that the microtubular complex becomes broken and ill-defined in transformed cells.

A striking rearrangement of the microtubular complex does occur at the onset of mitosis, the process of cell division. Immunofluorescence microscopy shows that as the chromosomes in the nucleus condense and begin to separate into two sets, the microtubules are broken down and the tubulin is reassembled to form the mitotic spindle: a framework of parallel microtubules that extends from the poles of the dividing cell to the chromosomes in the center and also stretches between the opposite poles. The spindle guides and probably supplies the driving force for the movement of the two sets of chromosomes to the poles. It is the cytoskeletal element that is re-

**WEB OF PROTEINS in the cytoplasm of a fibroblast, a connective-tissue cell, is made visible by fluorescence microscopy. Blue indicates the distribution of actin, the structural protein of microfilaments, which was stained with the fluorescent drug phalloidin. Red marks the distribution of vinculin, a protein that is associated with actin, and green shows the arrangement of tubulin, the structural protein of microtubules. The distribution of vinculin and tubulin was delineated immunologically: antibody to each protein was used to tag it with a fluorescent compound. J. Victor Small and Gottfried Rinnerthaler of the Institute of Molecular Biology in Austria constructed the image by making three separate black-and-white micrographs of the cell, in each of which only one protein was visible. The workers then passed the images through color filters and combined them photographically. The white areas are an artifact, and the yellow regions result from the superposition of colors.**

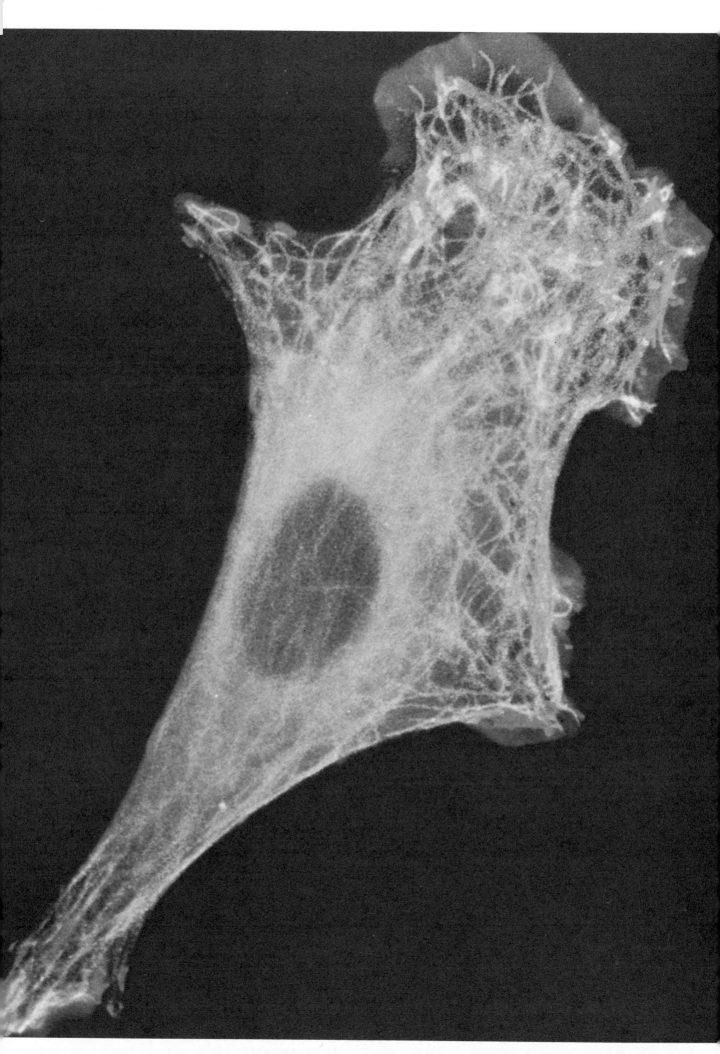

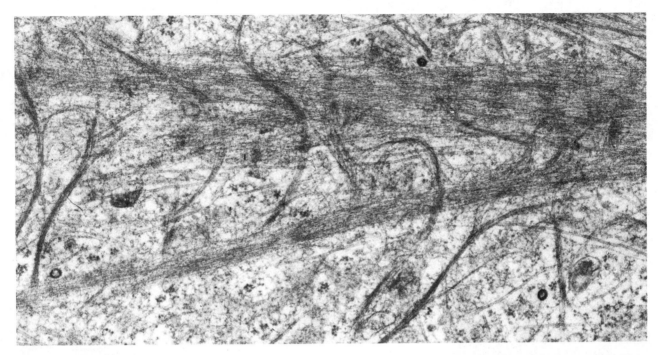

**PROTEIN FILAMENTS** lace the cytoplasm of an epithelial cell from a rat kangaroo, magnified 38,000 times. (Epithelial cells make up the surface layers of such tissues as skin and the lining of the intestine.) Diameter and appearance distinguish the three kinds of filaments in the cell matrix. The straight fibers that cross the electron micrograph in thick bundles are microfilaments. They are six nanometers (billionths of a meter) in diameter and are made of the protein actin. The bundles are called stress fibers. The isolated, thicker fibers, which resemble railroad tracks, are microtubules. They have a diameter of 22 nanometers and are composed of tubulin. The curved bundles are made up of intermediate filaments, which are between seven and 11 nanometers in diameter. Their protein varies among cell types; in epithelial cells they consist of keratin. David Henderson made the micrograph while working with the authors.

sponsible for the correct division of the parent cell's genetic material between the two daughter cells.

The antibody that is needed to observe the third filament system in a cell, the intermediate filaments, varies with their constituent protein, which in turn depends on the cell type. In keeping with their variable biochemistry, the arrangement of intermediate filaments is different in different tissues (and a few cells seem to lack them altogether). Intermediate filaments generally lace the entire cytoplasm, but in some cells they are bundled together and in others they are distributed as individual filaments. In many kinds of cells their arrangement appears to be related to that of the microtubules.

Immunofluorescence microscopy is a powerful tool for studying the overall arrangement of cytoskeletal filaments, and it has also proved valuable in determining which other proteins are associated with the different filament systems. With a few tricks the technique can even produce an image documenting three proteins in a single cell or stereoscopic images giving a three-dimensional impression.

Electron microscopy provides higher resolution than immunofluorescence microscopy. A section prepared for conventional electron microscopy is only about a two-hundredth the thickness of a cell, however, and therefore does not provide an overview of cytoskeletal organization. New high-voltage electron microscopes can be used to study thicker sections, in which individual cytoskeletal fibers can be followed for great distances. Another way to observe large-scale cytoskeletal organization with the electron microscope is to treat cells with a mild detergent. The detergent makes holes in the plasma membrane and extracts many soluble proteins from the cytoplasm, but it leaves the microfilaments, microtubules and intermediate filaments intact. Such preparations need not be sectioned before electron microscopy.

The configuration of the cytoskeletal elements can also be highlighted by quick-freeze deep-etch microscopy: quick-freezing and fracturing a cell, removing water from it and applying a thin layer of platinum. Under the electron microscope the treated specimen yields dramatic, seemingly three-dimensional images of the exposed cytoskeleton. Electron microscopy has also gained the biochemical specificity of immunofluorescence microscopy through the use of antibodies labeled with an electron-dense substance such as ferritin or colloidal gold; its distribution in a micrograph indicates the distribution of the target protein.

The techniques discussed above give only a static view of the cytoskeleton, a snapshot of the cell at the moment of its fixation. Other techniques have begun to reveal the behavior of the cytoskeleton in the living cell. One can chemically label a cytoskeletal protein in the test tube with a fluorescent compound and then inject it directly into a cell using a fine glass needle. The fluorescent tag makes it possible to follow the protein's incorporation into the cytoplasmic matrix with an image-intensifying video camera. To assess the role of a particular protein of the matrix in cell behavior and physiology one can inject antibody to the protein; often the antibody will inactivate the protein. The functioning of the injected cell will then give an indication of protein's role in normal cells.

To determine the molecular basis of the structure and behavior revealed by such microscopic techniques the observations must be combined with results from biochemical studies. In vitro assays in which cytoskeletal proteins are purified and their interactions with one another are observed have led to a biochemical description of much that is observed under the microscope. The proteins that underlie the organization and dynamics of the microfilaments, for example, are particularly well characterized.

Chief among them is actin, the basic protein of the microfilaments. In its purified form actin is a monomer (a single molecular unit that can combine with like units to form a polymer, or chain) known as globular actin, or G-actin. In the presence of adenosine triphosphate (ATP), a cellular energy-providing compound, and a physiological buffer the G-actin polymerizes to form long double helixes called filamentous actin, or F-actin. Simultaneously the ATP is converted into adenosine diphosphate (ADP), releasing the energy for the process. The F-actin chain has a polarity: it tends to polymerize, or elongate, at one end (the plus end) and to depolymerize, or shorten, at the other (the minus end). F-actin is the major structural element of the microfilaments, but in the cytoskeleton a number of other proteins, known as actin-binding proteins, link with it.

Actin-binding proteins govern the configuration and behavior of the F-actin. Analyses of skeletal muscle, smooth (involuntary) muscle and various nonmuscle cells and tissues have already documented several dozen such proteins, and more are sure to be found. Some are peculiar to specific cell types, but many are common to a range of cells and all are named for their in vitro effects on the assembly and structure of actin. Gelation factors markedly increase the viscosity of purified F-actin; they form flexible but tight links between crisscrossed filaments. Bundling factors, a set of very compact proteins, rigidly bind fibers in parallel, forming dense bundles that can be seen under the light microscope. Rodlike spacing factors form bridges between parallel actin filaments, linking them across distances of perhaps 200 nanometers.

Severing and stabilizing factors bind not to multiple actin strands but to single filaments, governing their length and stability. Severing factors seem to insert themselves between subunits in a filament, dividing it and forming a cap on the new plus end. When severing factors are added to a gel of F-actin, its viscosity decreases dramatically as filaments are cut. Stabilizing factors counter the action of severing factors; also known as cofilamentous proteins, they lie along the filament and thereby protect it. Tropomyosin, a short filamentous protein, may function as a stabilizing factor for F-actin in nonmuscle cells.

Still other proteins govern the assembly of individual F-actin fibers. Because actin fibers tend to add G-actin subunits at one end and to shed the monomer at the other, there is a flow of subunits through the chain when it is in equilibrium with free G-actin. The process is called treadmilling, and

it is regulated by capping factors. Each capping molecule is specific for either the plus or the minus end of the filament. Thus a capping factor halts either the addition of subunits to the filament end or their removal and thereby regulates the amount of actin in polymerized form.

Another set of actin-binding proteins that affect the equilibrium between G- and F-actin bind to the monomer rather than to the polymer.

These so-called sequestering factors form a complex with G-actin, preventing it from polymerizing. They provide a reservoir of monomeric actin, which the cell can tap when it needs to form new filaments.

How does a cell control the formation, linking, capping and severing of its microfilaments? At least some actin-binding proteins, notably the severing proteins and certain capping and spacing proteins, are regulated by flux-

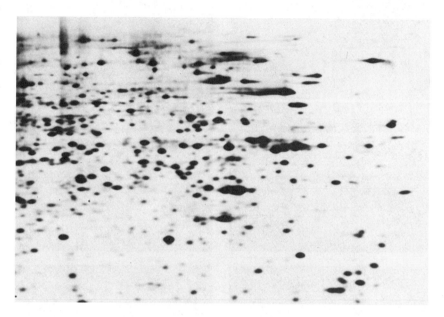

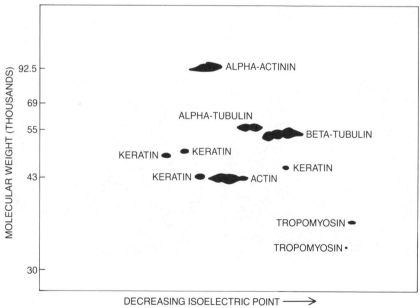

PROTEINS FROM EPITHELIAL CELLS were sorted first according to charge, here expressed as isoelectric point (*left to right*), and then according to their molecular weight (*top to bottom*) by two-dimensional gel electrophoresis. Living cells were exposed to a mixture of radioactively labeled amino acids, which the cells gradually incorporated into their proteins. The proteins were separated by electrophoresis and a photographic emulsion was placed on the gel; radioactivity from the proteins produced the image at the top. The cytoskeletal proteins are major components of the cell. Actin, tubulin and keratin, identified at the bottom, are the basic proteins respectively of microfilaments, microtubules and (in epithelial cells) intermediate filaments. The other identified proteins are linked with the three basic proteins. R. Bravo and J. E. Celis of the University of Aarhus supplied the image.

es of the calcium ion, which is known to be an important cellular messenger. The sequestering factors may respond to certain lipids; in vitro the lipids cause sequestering factors to release their complement of bound G-actin. It is quite likely that some cells also con-

trol their microfilament architecture by altering the relative proportions of the various actin-binding proteins.

Actin-binding proteins influence not only the organization of the actin mesh but also its activity. The capping and severing proteins, for example, may

play a part in some kinds of cell locomotion and in movement within the cell by controlling the local shortening and lengthening of actin filaments. The acknowledged motor of cell movement and possibly of certain aspects of intracellular transport is yet another actin-binding protein, myosin. Together with actin myosin can convert the chemical energy of ATP into contractile movement. Myosin is in fact an ATPase, an enzyme that breaks down ATP into ADP. Its activity is stimulated when it is bound to F-actin.

The association of myosin with F-actin is well established in muscle cells, where the two proteins form orderly contractile units known as sarcomeres. The cytoplasm of nonmuscle cells contains a much smaller amount of myosin in a far less ordered state. Instead of forming the thick, bipolar (double-headed) filaments typical of muscle, the myosin is morphologically lost in the jungle of cytoskeletal organization that is visible in ordinary electron micrographs. Immunofluorescence microscopy, however, confirms its presence among the microfilaments.

Myosin's importance in contraction can be demonstrated in vitro, and the probable biochemistry of the process has been unraveled. Once again a flux of calcium ions acts as a trigger. Myosin kinase, another protein associated with microfilaments, together with calmodulin, a substance that mediates many effects of calcium, alters the configuration of the myosin molecules. They stretch and aggregate into small bipolar filaments. The calcium ions also activate severing factors, which cut areas of the actin network. The bipolar myosin filaments then bind at both ends to properly aligned actin filaments and, in the presence of ATP, slide the filaments along each other. It is also possible that myosin molecules, appropriately anchored, could tug on single actin filaments.

Direct proof that F-actin and myosin together convert the chemical energy stored as ATP into movement comes from a recent experiment done with the purified proteins. A system of actin filaments running in parallel and having identical polarity was attached to a carbon film. Small polystyrene beads to which myosin molecules had been linked were then placed on the actin. When ATP was added to the system, the beads were observed to move along the filaments. The rate of movement was consistent with the speed of muscle contraction and of certain forms of cell motion. Just how a living cell translates such movement at the molecular level into cell locomotion and the intracellular movement of or-

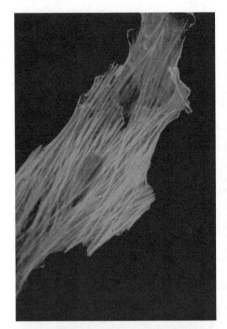
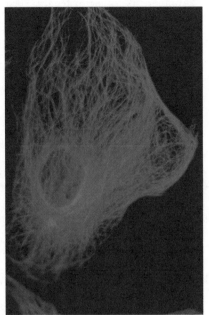
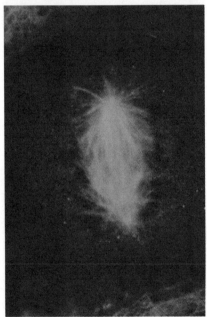
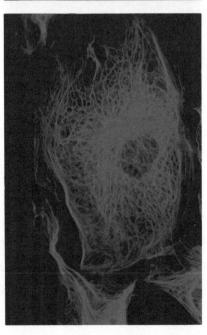

**CYTOSKELETAL FILAMENT SYSTEMS** have **distinct configurations that are evident in immunofluorescence micrographs. Microfilaments (top left) parallel the cell-surface membrane in bundles known as stress fibers, which may have a contractile function. Just below the membrane microfilaments also form a fine mesh, faintly evident here at the cell's ruffled upper border, which marked its leading edge while it was living and mobile. Microtubules (top right), visible individually in this micrograph, radiate from an area near the nucleus known as the cytocenter (most heavily stained region). Microtubules may act as radial guides for intracellular transport in the nondividing cell, but for cell division they break down and the tubulin is reassembled into the microtubules of the mitotic spindle, shown in a micrograph for which a different fluorescent compound was used (bottom left). Intermediate filaments (bottom right), like microtubules, extend throughout the cytoplasm. In an epithelial cell, such as the one shown, intermediate filaments form thick, wavy bundles that probably provide structural reinforcement; in other types of cells the intermediate filaments usually are not as tightly aggregated as in epithelium and their function is less clear.**

ganelles is not understood, however.

The relation of the microfilament network to the architecture of the cell is better known than the network's role in movement. Microfilaments dominate the outer part of the cytoplasm; they therefore have a strong influence on the characteristics of the cell surface. A specialized actin-based scaffolding underlies the plasma membrane of some cells. In certain cells it has been possible to purify and isolate the entire scaffolding. Immunological and biochemical techniques have revealed the biochemical makeup of the isolated protein structure.

Just under the plasma membrane of red blood cells very short actin filaments and an actin-binding protein form a two-dimensional mesh that seems to account for the cells' flexibility, a feature that enables them to pass through extremely fine capillaries. The actin chains, which probably contain only about a dozen actin monomers, lie at the junctions of the network; they are interconnected by long molecules

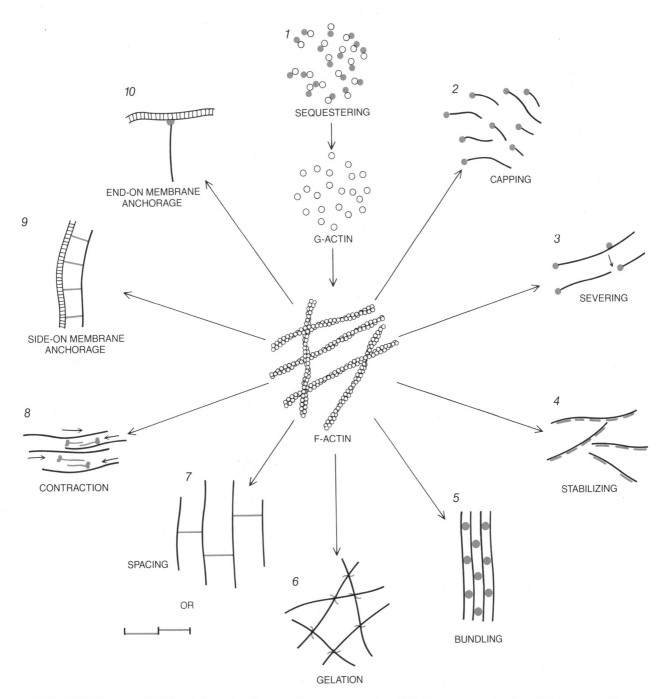

**ACTIN-BINDING PROTEINS** (*color*) govern the organization and activity of actin (*black*), the basic protein of the microfilaments. Globular actin (G-actin), the protein subunit of the microfilaments, is held in storage by profilin, a protein that acts as a sequestering factor (*1*). When it is released, the G-actin polymerizes to form helical strands known as filamentous actin (F-actin). F-actin strands have a polarity: they tend to add molecules of G-actin at one end and shed them at the other. Capping factors (*2*), by binding to one or the other end of the strand, regulate the process. Sever-ing factors (*3*) divide completed strands; stabilizing factors (*4*) counteract severing factors by binding in parallel to actin strands, thereby protecting them. Bundling (*5*), gelation (*6*) and spacing (*7*) factors dictate the spatial organization of actin filaments; the contraction-producing factor, myosin (*8*), makes the organization a dynamic one. It binds to actin filaments having opposite polarity (*arrows*) and slides them along each other. The process requires the energy-carrying compound adenosine triphosphate (ATP). Still other proteins (*9, 10*) anchor the network to the cell-surface membrane.

of a spacing factor, spectrin. A third protein, ankyrin, binds the spectrin cross-links to a major transmembrane protein (a protein that spans the plasma membrane) and anchors the network; other proteins further stabilize the basic architecture.

The exposed surface of an intestinal epithelial cell has a more complex topography, and a more intricate cytoskeletal scaffolding underlies it. A single cell carries some 1,000 fingerlike projections known as microvilli, which add to the area of surface through which nutrients are absorbed. The microvilli are loaded with digestive enzymes and biochemical transport systems that process the nutrients. Each microvillus is stabilized by a bundle of F-actin, in which the proteins villin and fimbrin act as bundling factors. A series of helically arranged bridges link the bundle to the plasma membrane of the microvillus. The protein of the bridges has not been characterized completely, but it may be a transmembrane protein that extends inward to meet the actin bundle. At the base of the bundle the filaments are rooted in a horizontal net that is thought to be dominated by a spectrinlike spacing factor. The net holds the many core filaments in place and thereby stabilizes the microvilli.

Electron micrographs have shown many other kinds of interaction between microfilaments and cell surface membranes. Many such interactions are certain to have structural significance, as they do in red blood cells and the intestinal epithelium; the molecular details of the interaction are not as well understood in other cells.

It is also likely that the microfilament net under the plasma membrane has an important bearing on cell physiology. In the red blood cell the transmembrane protein, known as band III for its position on electrophoretic gels, provides an anchorage for certain glycolytic enzymes, which mediate the breakdown of glucose. Band III also acts as a channel allowing ion exchange across the membrane, enabling the red cells to discharge carbon dioxide as they pass through the lung tissue, where they also acquire oxygen. In other cell types certain protein kinases (substances that play a regulatory role in cell physiology) are present in the subsurface mesh. Moreover, signals received on the outer surface of the membrane by receptors that bind messenger molecules may directly affect the subsurface net.

Certain tumor viruses may also act on the subsurface mesh. In transformed cells proteins encoded by the oncogenes (cancer-inducing genes) of the viruses may be localized within the mesh. Such proteins can cause dramatic changes in the cytoskeleton; in some instances they have been shown to act by phosphorylating (adding a phosphate group to) specific cytoskeletal proteins.

The second filament system, the microtubules, differs from the microfilaments in overall organization, as immunofluorescence micrographs show. It also has a different set of functions. It appears to be a major factor in intracellular organization and transport as well as in the architecture of the cell as a whole. Microtubules resemble microfilaments, however, in having a single major structural protein. A substance called tubulin is the stuff of microtubules and also of related organelles such as cilia and flagella (two kinds of appendages with which certain cells propel themselves or move surrounding liquid or particles), the basal bodies to which cilia and flagella are attached and the centrioles.

Like the actin of microfilaments, tubulin forms polymers. Tubulin itself is a globular protein that consists of two distinct but similar polypeptides, alpha-tubulin and beta-tubulin. The polymerization of tubulin can be ob-

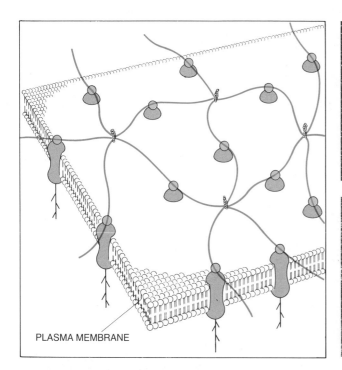

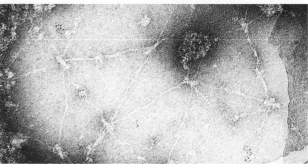

FRAMEWORK OF ACTIN AND SPECTRIN on the inner surface of the membrane of a human red blood cell gives the cell strength and flexibility. Very short actin chains (*black in diagram at left*) form the "knots" of the net; long, thin molecules of the spacing factor spectrin (*color*) interconnect the actin. The spectrin binds to a protein known as ankyrin (*light color*), which fastens the framework to a protein known as band III (*gray*) that is embedded in the membrane and protrudes from the outer surface. Other protein components (not shown) further stabilize and strengthen the subsurface structure. An electron micrograph of a purified spectrinlike protein (*top right*), made by John Glenney while he was working with the authors, shows its elongated shape. A micrograph of red-cell membrane (*bottom right*), which was stretched to about 10 times its original size to expose the organization, reveals spectrin links, the ankyrin-binding sites on the links and the actin chains. Timothy J. Byers and Daniel Branton of Harvard University provided the micrograph.

served in vitro. In the presence of guanosine triphosphate (GTP), an energy-carrying compound similar to ATP, the tubulin dimers (the two-part molecules made up of alpha- and beta-tubulin) join to form a tube with a hollow core. (It is remarkable that some 20 years after the structure of microtubules was observed through the electron microscope it is still not known whether the central space, which is about 10 nanometers in diameter, has any function.)

Microscopic and biochemical studies suggest that microtubules, like microfilaments, are associated with other proteins that influence their organization and activity. Electron microscopy shows that a number of associated proteins project from the wall of microtubules into the cytoplasm, sometimes for tens of nanometers; their function may be to link the microtubules with other cytoplasmic components or to define a protected zone around each microtubule.

The biochemical properties of tubulin point to the existence of other associated proteins. Like F-actin, microtubules display a polarity in vitro: one end tends to extend by polymerization whereas the other tends to shorten, leading a process of treadmilling similar to the one demonstrated for F-actin. Thus certain proteins or more complex substances may act as capping factors, exerting pronounced effects on the formation and breakdown of microtubules. In vitro, centrosomes appear to nucleate the assembly of microtubules that grow and shrink. Their dynamic instability may explain certain aspects of microtubule behavior in vivo, particularly in the mitotic spindle. Interestingly, microtubule ends containing GTP (the energy-rich form) seemed more resistant to shortening by depolymerization than ends with a larger proportion of GDP.

Some tubulin-associated proteins may be present only at specific stages in a cell's existence or in specific compartments of the cell, which suggests that they may have unique functions. There are now tantalizing indications, for example, of associated proteins that are unique to the tubulin of the mitotic spindle. In neurons different proteins seem to be associated with microtubules in the dendrites and in the axons. In cilia and flagella tubulin links with a flexible protein known as dynein. Like myosin, dynein is an ATPase, and it powers the beating motion of cilia and flagella.

Dynein or some other ATP-dependent analogue of myosin may bind to tubulin in the cytoplasm too. If such an association did occur, a cytoplasmic system of tubulin and dynein might provide a further basis for cell movement. Tubulin might also support some movement through capping factors and the shortening or lengthening of microtubules.

A different mechanism based on tubulin and a putative tubulin-associated protein may transport vesicles along the axons of nerve cells and perhaps move other organelles within the cytoplasm. A recent in vitro study showed that vesicles isolated from the giant axons of the squid move along purified microtubules at speeds of up to one micrometer per second when ATP and a crude mixture of cellular proteins are added. It is thought that a protein in the mixture, unidentified as yet, acts as a translocator, probably binding both to vesicles and to a microtubule. The hypothetical translocator would act as an ATPase, breaking down ATP to derive the energy needed for vesicle translocation.

The details of microtubular dynamics in the cytoplasm remain uncertain; evidence for their structural roles is stronger. The polarities of all the microtubules radiating from the centrosome are probably identical. If a cell is treated with the drug colcemid, its microtubules depolymerize and essentially disappear. When the drug is washed away, the microtubules reappear in a strikingly unidirectional manner, elongating outward from the centrosome at a rate of about one micrometer per minute (at a temperature of 37 degrees Celsius), which is about the same rate that is observed when tubulin polymerizes in vitro. Within 75 minutes a microtubular display similar to that of a normal cell has reappeared. The directional nature of

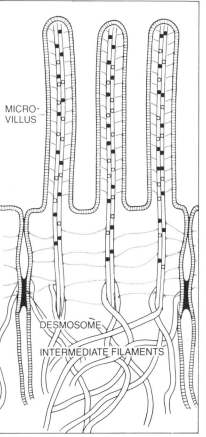
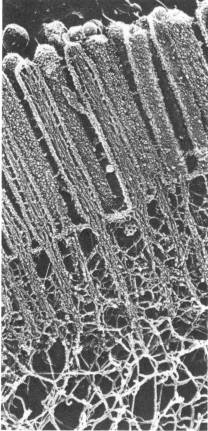

**ACTIN-BASED SCAFFOLDING** stiffens the microvilli (fingerlike projections) on the surface of an intestinal epithelial cell. Within each microvillus the bundling factors villin (*circles in diagram at left*) and fimbrin (*dots*) bind actin filaments (*black*) in a longitudinal bundle; another protein (*color*) forms lateral bridges between the bundle and the surface membrane of the microvillus. It is not known how the bundle is anchored at its upper end. Below the microvilli a variant of the spacing factor spectrin (*gray*) forms a mesh in which the bundles are rooted. Deeper still the actin fibers intermingle with the cell's intermediate filaments, which are made up of keratin. The keratin filaments in turn are anchored at the cell membrane at specialized regions known as desmosomes, which cement together epithelial cells. A deep-etch micrograph (*right*) of an intestinal cell that had been treated with detergent to remove soluble proteins shows a number of microvilli. In several of them actin bundles and their connections with membrane can be seen, as can the spectrin and keratin webs. John E. Heuser of Washington University in St. Louis provided the micrograph.

microtubules suggests they could act as guides for directed radial transport within the cell.

The microtubules also seem to play a part in establishing the distinctive shape and orientation of certain cells. A fibroblast (a connective-tissue cell), for example, is asymmetric, displaying a ruffled leading edge and a slender tail as it moves across a substrate. When such a cell is treated with colcemid, causing the microtubules to disappear, it becomes more symmetrical and its locomotion stops.

Such a treatment also visibly disturbs the workings of the cell. The characteristic motion of intracellular organelles and certain vesicles—intermittent, directed movements—stops. The membranes of the Golgi apparatus become disordered and move away from their normal location near the centrosome, and the interconnected system of channels making up the endoplasmic reticulum seems to retract. Disturbing the microtubules in nonepithelial cells also seems to affect another system of fibers, the intermediate filaments: they retract into the interior of the cell and coil around the nucleus.

Experiments with other drugs that disturb the normal arrangement of microtubules have yielded further evidence that microtubules define spatial order within the cell. Taxol, a drug that promotes rather than inhibits the polymerization of tubulin, causes a cell's microtubules to form bundles that no longer connect to the centrosome; similar effects result from the injection of tubulin-specific antibodies into a cell.

Microtubules thus appear to establish the geometry of the cell, acting as tracks that orient other cellular phenomena. In cells of limited dimensions such guidance might not be needed, at least for intracellular transport: the vagaries of Brownian motion might suffice to deliver vesicles and other organelles to their proper sites within the cell. In highly asymmetric cells such as neurons, however, whose axons extend for several meters in some animals, guidance may be essential to transport within the cell.

Less is known about the function of the intermediate filaments, the third element of the cytoskeleton. It is also harder to generalize about their biochemistry. The basic proteins of intermediate filaments are encoded by a single family of genes, but different genes of the family are expressed in different kinds of cells and tissues. By the late 1970's immunological and biochemical studies had shown that the genes for intermediate-filament proteins are expressed in accordance with the path the tissue followed during embryogenesis.

The diversity of paths gives rise to five biochemically distinct kinds of intermediate filaments: epithelial keratins in epithelial cells, neurofilaments in most (but probably not all) neurons, desmin filaments in muscle, glial filaments in glial cells (the supporting cells in the brain, the spinal cord and the peripheral nervous system) and vimentin filaments in cells of mesenchymal origin, such as those in connective tissue and in blood and lymph vessels. Subtypes exist within the epithelial keratins; some 20 different molecules have been catalogued in human tissue, some of them specific to morphologically distinct epithelia.

The molecules of intermediate-filament proteins all have a rodlike central region of unvarying length. Within this region the proteins are very similar: between 30 and 70 percent of the constituent amino acids are identical. The rodlike region is the structural basis of intermediate-filament assembly. A cross section through a single 10-nanometer filament would show about 30 molecules in an interlocking arrangement. In contrast to the polymerization of microfilaments and microtubules from globular subunits, the assembly of intermediate filaments does not require ATP or GTP.

The terminal regions of the intermediate-filament proteins are far more variable in size and in amino acid makeup than the central rods. In general the terminal regions do not take part in binding the molecules into a filament. They probably extend from the filament into the cytoplasm, where they may function to link the filament with other structural components. One end of the largest neurofilament protein, for example, accounts for the wispy cross bridges that are seen joining neighboring axonal neurofilaments in electron micrographs.

Like the other filament systems the intermediate filaments have a set of associated proteins, only some of which are now known. Their role, and indeed the function of the entire network, remains rather elusive. When antibodies to intermediate-filament proteins are injected into cultured cells, cell locomotion and cell division do not appear to be affected. Intermediate filaments may play a subtle role in cell dynamics that is difficult to identify in isolated cells grown in a laboratory culture.

In some cells and tissues, however, it is apparent that the intermediate filaments have a structural purpose. It is widely believed intermediate filaments are attached at one end to the membrane that envelops the nucleus, and it is known that in epithelial cells many of the keratin filaments are anchored at specialized regions of the cell surface membrane known as desmosomes. Desmosomes are a kind of intercellular junction that serves to join neighboring cells together. At a desmosome specialized proteins on each cell extend through the membrane and project a glycoprotein end at the outer surface; the glycoproteins interact with similar ones on the other half of the desmosome to unite the cells. Together the keratin fibers and the desmosomes integrate the epithelium, giving it stability and strength. In some nonepithelial cells the intermediate filaments may link instead with the actin-spectrin web under the membrane.

The three filament systems are not completely separate and independent units of the cell architecture. The similarities in the arrangement of mi-

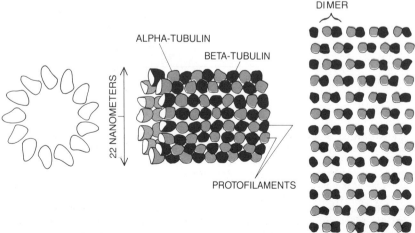

**ASSEMBLY OF TUBULIN** takes place as dimers made up of alpha- and beta-tubulin link to form 13 protofilaments that are arrayed around a hollow core. The result is a microtubule **22 nanometers in diameter, shown end on (*left*), in a side view (*middle*) and unrolled (*right*).**

crotubules and intermediate filaments and the effects on the intermediate filaments of nonepithelial cells when the microtubules are disrupted suggest that the two systems are linked, probably by associated proteins. It also seems likely that microtubules may act as the scaffolding on which the more permanent framework of intermediate filaments is erected. Electron micrographs hint at connections between microfilaments and intermediate filaments as well. Indeed, the notion that the three filament systems form an integral framework with which organelles, and perhaps even enzymes and soluble proteins, may associate is widely discussed.

Some investigators have proposed that a fourth fibrous system integrates the other filament systems. This "microtrabecular lattice," whose existence was implied in certain high-voltage electron micrographs, is held to be an irregular mesh of fine fibers whose diameter ranges from less than two nanometers to more than 10; all the other fibrous elements in the cell, as well as its organelles, would be embedded in the microtrabeculae, which would act as the cell's ground substance. The protein makeup of the microtrabeculae presumably would differ from that of the other three systems, but the existence of an additional major cellular protein that has not yet been identified in biochemical analyses of the cytoplasm is rather unlikely.

Instead of being a fourth system the lattice seen in the micrographs probably reflects the intricate interconnections and interdigitations of the three established systems as well as some artificial fraying of the cytoskeletal structures. It is also possible that during the preparation of specimens for high-voltage electron microscopy proteins dissolved in the cytoplasm become fixed onto the filamentous arrays, increasing their apparent complexity. Indeed, improvements in fixing techniques have recently resulted in high-voltage electon micrographs in which the microtrabecular system is essentially absent.

Just as it is not yet possible to describe the cell matrix as a single structure, so an overall understanding of its role in cell dynamics is emerging only slowly. Microfilaments are crucial to cell locomotion and surface movement and microtubules are crucial to mitosis, but the molecular intricacies of their roles are largely unknown. Indeed, it is often unclear whether the matrix is necessary for, or only enhances, a particular cellular function. It is possible to argue, for example, that the process fundamental to

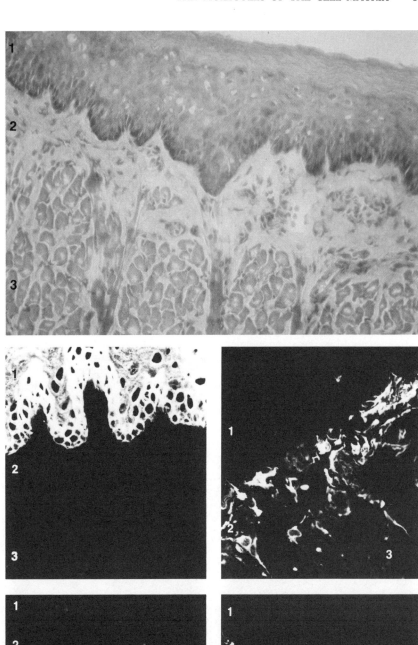

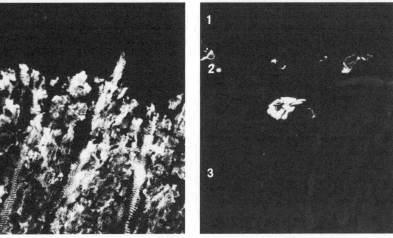

CELL TYPES can be distinguished in complex tissues by the intermediate-filament protein they contain. The micrographs show comparable sections of tongue; the lower four images were made using fluorescent-labeled antibodies to different intermediate-filament proteins. The top micrograph, shown for comparison, was stained conventionally, with hematoxylin eosin. The numbers identify the same layer of tissue in different micrographs. Antibody to keratin binds to and makes visible the epidermal cells of layer 1 (*center left*); antibody to vimentin distinguishes the cells of connective tissue and blood vessels, concentrated in layer 2 (*center right*); antibody to desmin reveals the muscle cells that make up layer 3 (*bottom left*), and antibody to the neurofilament protein highlights the axons and neuronal cell bodies that are distributed throughout the section of tongue tissue (*bottom right*).

cell locomotion is the extension of the advancing edge of the cell through the addition of membrane there; the myosin-mediated contraction of the microfilaments would then serve merely as the follow-up mechanism needed to pull along the rest of the cell behind the advancing membrane.

The techniques of molecular genetics may help to clarify the function of cytoskeletal proteins. It has already been shown that certain mutations in the gene encoding actin in yeast disturb the synthesis, assembly and integrity of the plasma membrane and the movement of secretory vesicles to the cell surface; the finding implies that microfilaments play a role in both processes. The creation of mutations at specified sites in the genes coding for the cytoskeletal elements and the introduction into the cell of cloned DNA coding for altered cytoskeletal proteins should make it possible to observe the effects not only of a single alteration in the matrix but also of multiple changes. Such techniques should elucidate structural and dynamic interrelations among the cytoskeletal proteins.

Much remains to be learned about the cell matrix. What is known has already contributed to diagnosis and research in human pathology. Because certain proteins of the matrix, and in particular those of the intermediate filaments, vary from tissue to tissue, the proteins provide a basis for distinguishing cell types. The ability to classify cells correctly is crucial to the diagnosis of cancers, which can metastasize to sites far from the tissue in which they originated. Appropriate treatment can depend on knowing the cellular origin of the tumor.

Typing of intermediate filaments through immunofluorescence microscopy distinguishes the major tumor groups. Thus keratins are found in carcinomas (tumors of epithelial origin), the glial-filament protein in tumors of glial origin, desmin in muscle-cell sarcomas, vimentin in lymphomas and nonmuscle sarcomas, and the proteins of neurofilaments in tumors originating in the sympathetic nervous system. In the case of carcinomas the many variant forms of keratin make possible finer diagnostic distinctions:

between squamous-cell carcinomas and adenocarcinomas, for example. The technique can enable pathologists to characterize rapidly and unambiguously some of the 5 to 10 percent of tumors that are difficult to diagnose using conventional pathological stains. Because the method is extremely sensitive, it can also be used to detect micrometastases of a few tumor cells to lymph nodes or to the bone marrow.

Intermediate-filament typing has other medical applications. Combined with amniocentesis it can reveal certain congenital malformations. The presence in the amniotic fluid of cells containing glial filaments or neurofilaments, for example, can indicate a fetus with a malformation of the central nervous system. The technique has also revealed intermediate-filament abnormalities in the muscle of patients suffering from certain disorders of heart and skeletal muscle, in the liver of alcoholics and possibly in the brain of people with Alzheimer's disease. Such applications illustrate how basic research in cell biology can contribute to the diagnosis and understanding of human disease.

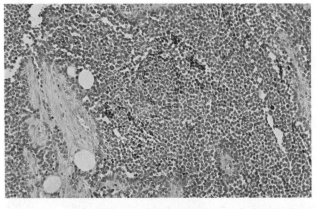
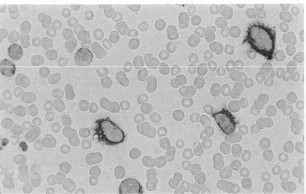
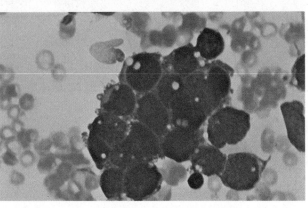
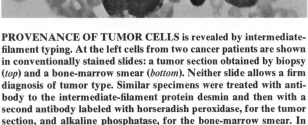

**PROVENANCE OF TUMOR CELLS is revealed by intermediate-filament typing. At the left cells from two cancer patients are shown in conventionally stained slides: a tumor section obtained by biopsy (*top*) and a bone-marrow smear (*bottom*). Neither slide allows a firm diagnosis of tumor type. Similar specimens were treated with antibody to the intermediate-filament protein desmin and then with a second antibody labeled with horseradish peroxidase, for the tumor section, and alkaline phosphatase, for the bone-marrow smear. In** both slides the immunologic probe strongly stained the cytoplasm of tumor cells (*brown cells at top right, red cells at bottom right*). Desmin is the intermediate-filament protein that is characteristic of muscle; its detection made it possible to identify the tumor in both cases as a rhabdomyosarcoma, a variety of malignancy that originates in skeletal muscle. The micrographs at the top were made in the authors' laboratory in Göttingen; David Mason of the University of Oxford provided the micrographs that are shown at the bottom.

# 7

# THE MOLECULES OF THE IMMUNE SYSTEM

# The Molecules of the Immune System

*The proteins that recognize foreign invaders are the most diverse proteins known. They are encoded by hundreds of scattered gene fragments, which can be combined in millions or billions of ways*

by Susumu Tonegawa

The immune system is clearly essential to survival; without it death from infection is all but inevitable. Even apart from its vital function the immune system is a fascinating example of biological ingenuity. The cells and molecules of this defensive network maintain constant surveillance for infecting organisms. They recognize an almost limitless variety of foreign cells and substances, distinguishing them from those native to the body itself. When a pathogen enters the body, they detect it and mobilize to eliminate it. They "remember" each infection, so that a second exposure to the same organism is dealt with more efficiently. Furthermore, they do all this on a quite small defense budget, demanding only a moderate share of the genome and of the body's resources.

The critical event in mounting an immune response is the recognition of chemical markers that distinguish self from nonself. The molecules entrusted with this task are proteins whose most intriguing property is their variability of structure. In general all the molecules of a given protein made by an individual are absolutely identical: they have the same sequence of amino acids. At most there may be two versions of a protein, specified by maternal and paternal genes. The recognition proteins of the immune system, in contrast, come in millions or perhaps billions of slightly different forms. The differences enable each molecule to recognize a specific target pattern.

The most familiar of the recognition proteins are the antibodies, or immunoglobulins. Much has been learned of their structure and, equally important, of the genetic mechanisms responsible for their diversity. It turns out that vast numbers of antibodies are made by reshuffling a much smaller set of gene fragments. Thus antibody genes offer dramatic evidence that DNA is not an inert archive but can be altered during the life span of an individual. In the synthesis of antibodies the cutting and joining of gene sequences is not a mere incidental feature of the genetic process; it is essential to the functioning of the immune system.

Another class of recognition molecules consists of the proteins called *T*-cell receptors. Because they are more difficult to isolate, their properties are not yet as well known as those of the antibodies. In structure and evolutionary origins they are clearly related to the immunoglobulins, and a similar genetic mechanism accounts for their diversity, but their mode of operation is subtly different. A *T*-cell receptor recognizes only those cells that bear both self and nonself markers. By this curious means *T* cells are given the ability both to act directly against viral infections and to regulate other components of the immune system.

The primary cells of the immune system are the small white blood cells called lymphocytes. Like other blood cells, they are derived from stem cells in the bone marrow. In mammals one class of lymphocytes, the *B* cells, complete their maturation in the bone marrow. A second class, the *T* cells, undergo further differentiation in the thymus gland. Cells of the two classes are similar in size and appearance, but they take part in different forms of immune response.

*B* lymphocytes are the cells that manufacture antibodies. Their basic mode of action can be understood in terms of the clonal selection theory proposed 30 years ago by Sir Macfarlane Burnet. As each *B* cell matures in the bone marrow, it becomes committed to the synthesis of antibodies that recognize a specific antigen, or molecular pattern. In the simplest case all the descendants of each such cell retain the same specificity, and so they form a clone of immunologically identical cells. (Actually some variation is introduced as the cells proliferate.)

The antibodies made by a *B* cell remain bound to the cell membrane, where they are displayed on the surface as receptor molecules. When an antigen binds to an antibody in the membrane, the cell is stimulated to proliferate; this is the clonal-selection process. In general many clones respond to a single infection. The antigenic markers recognized by antibodies are comparatively small patterns of molecular structure, and a single virus

BINDING OF AN ANTIGEN to an antibody is a central event in the recognition of foreign organisms in the body. In the computer-generated image on the opposite page the bound substance is not an actual antigen but a hapten, a small molecule with an affinity for a particular antibody. The hapten shown is phosphocholine. It is guided to the antigen-binding site by electrostatic interactions and fits into a cleft on the antibody surface. Its orientation on approaching the binding site, as depicted in the top center of the image, is suggested by a calculation done by Elizabeth D. Getzoff, John A. Tainer and Arthur J. Olson of the Research Institute of Scripps Clinic; the calculation is based on the atomic structure of the antibody-hapten complex, which was determined by Eduardo A. Padlan, Gerson H. Cohen and David R. Davies of the National Institutes of Health. The skeleton of the protein and that of the approaching hapten molecule are shown enveloped in dots that represent the surfaces accessible to water molecules; another hapten, installed in the antigen-binding site directly below the first hapten, is shown as a skeleton only. The colors of the dots indicate the calculated electrostatic potential of various regions of the molecular surface; blue is the most positive and red the most negative. Arrows show the direction of the electrostatic field, and their colors indicate the electrostatic potential at their points of origin. The image was made with the programs GRAMPS, developed by Olson and T. J. O'Donnell of Abbott Laboratories, and GRANNY, written by Olson and Michael L. Connolly of Scripps Clinic.

or bacterium carries many recognizable patterns.

Some of the progeny of the selected clones remain as circulating *B* lymphocytes. They serve as the immune system's memory, providing a faster response to any subsequent exposure to the same antigens. The memory cells are responsible for the immunity that develops following many infections or as a result of vaccination. Other members of the selected *B*-cell clones undergo "terminal differentiation": they grow larger, stop reproducing and dedicate all their resources to the production of antibodies. In this state they are called plasma cells, and although they live for only a few days more, they secrete large quantities of immunoglobulins.

Antibody molecules cannot destroy a foreign organism directly; they mark it for destruction by other defensive systems. One of these systems is complement, a set of more than a dozen proteins that are activated in succession on the surface of a cell bearing antigen-antibody complexes. The complement proteins have the ultimate effect of perforating the cell membrane. Antigen-antibody complexes also attract macrophages, which engulf and digest foreign particles. A number of other cells can also take part in the immune response.

How does an antibody molecule

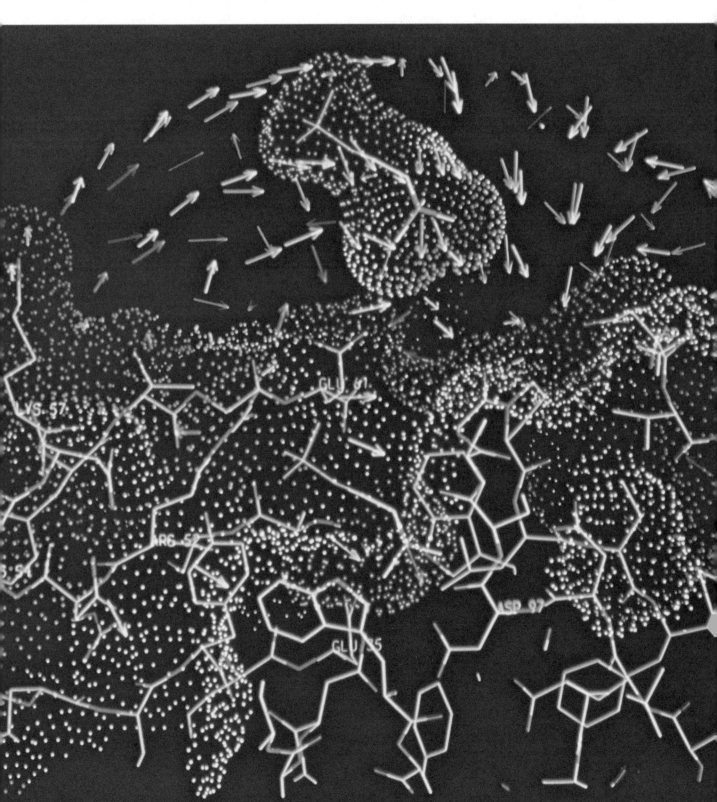

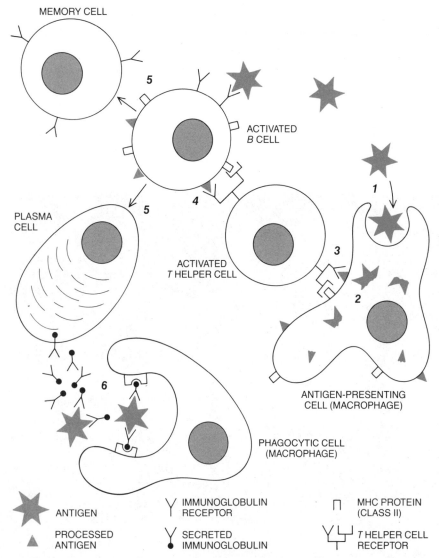

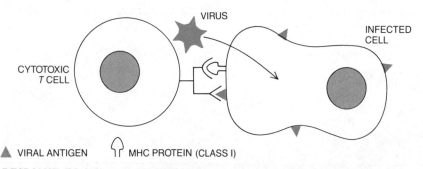

IMMUNE RESPONSE TO INFECTION mobilizes several cooperating populations of cells. B cells carry immunoglobulins as surface receptors that recognize and bind to circulating antigens; in general, however, the B cells are not activated by this process alone. First the antigen must be taken up by an antigen-presenting cell (1); a macrophage can serve in this role. The antigen is processed by the macrophage (2) and then displayed on its surface. There it is recognized by a T helper cell, which is thereby activated (3). The T helper cell then activates B cells carrying the same processed antigen (4). The activated B cells proliferate and undergo terminal differentiation (5). Some of the progeny become memory cells, which provide a quicker response to future infections, whereas others develop into antibody-secreting plasma cells. The secreted antibodies bind to the antigen, thereby marking it for destruction by various other components of the immune system, including macrophages (6).

RESPONSE TO VIRAL INFECTION calls on other elements of the immune system. When a virus enters a cell, viral proteins are left behind embedded in the cell membrane. Cytotoxic T cells specifically recognize such foreign molecules displayed in combination with proteins that identify the host. They are the Class I proteins of the major histocompatibility complex (MHC). The cell infected by the virus is killed by the cytotoxic T cell.

recognize an antigen? The answer was found by analyzing the amino acid sequence and the three-dimensional structure of antibodies.

A basic antibody molecule consists of four polypeptide strands: there are two identical light chains of about 220 amino acids and two identical heavy chains of either 330 or 440 amino acids. The four chains are held together by disulfide bonds and noncovalent bonds to form a Y-shaped molecule. Both the heavy and the light chains are built up from a common domain, or structural subunit, of about 110 amino acids. It would appear that the gene for some prototypical protein of about this size has been repeatedly duplicated and altered to give rise to the genes for both immunoglobulin chains. A light chain has two somewhat different copies of the domain and a heavy chain has either three or four copies. All the copies fold up into broadly similar three-dimensional structures.

In both heavy and light chains the domain at the amino end of the polypeptide—the end synthesized first—differs in an important way from the other domains. The amino-terminal domain is where most of the variation in amino acid sequence is found. In the folded antibody the variable regions make up the terminal half of the arms of the Y; each arm incorporates the variable region of one heavy chain and one light chain. Within the variable region of each chain there are three small segments where the amino acid sequences are found to be particularly diverse. These "hypervariable" segments come together at the end of each arm to form a cleft that acts as the antigen-binding site. The specificity of the molecule depends on the shape of the cleft and on the properties of the chemical groups that line its walls; thus the nature of the antigen recognized by an antibody is determined primarily by the sequence of amino acids in the hypervariable regions.

One further aspect of the structure of antibodies must be mentioned. Even in the constant regions not all molecules are identical. In mammals the light-chain constant regions are of two types, designated kappa and lambda. There are five classes of heavy-chain constant regions: mu, delta, gamma, epsilon and alpha. Antibodies with the same variable regions but different heavy-chain classes recognize the same antigens but have different roles in the immune response. For example, the membrane-bound antibodies that serve as B-cell receptors incorporate mu or delta chains, and most of the antibodies secreted in response to an an-

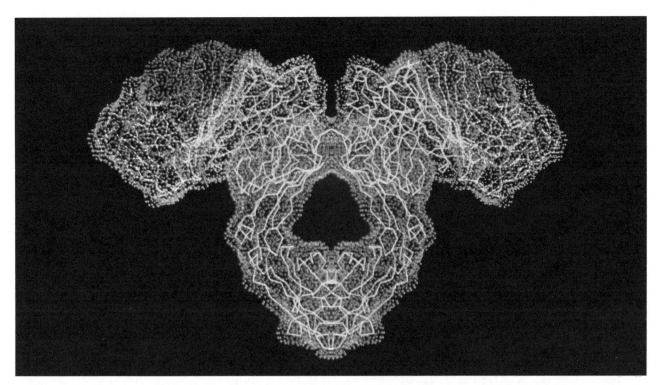

**ANTIBODY MOLECULE** is a Y-shaped protein made up of four polypeptide chains. Two heavy chains (*blue surfaces*) extend from the stem of the Y into the arms; two light chains (*green surfaces*) are confined to the arms. Each polypeptide has both constant regions (*white and yellow skeleton*) and variable regions (*red skeleton*). All antibodies of a given type have the same constant regions, but the variable regions differ from one clone of *B* cells to another. At the end of each arm the heavy- and light-chain variable regions fold to create an antigen-binding site. The image was made by Olson with the computer programs used in making the one on page 73.

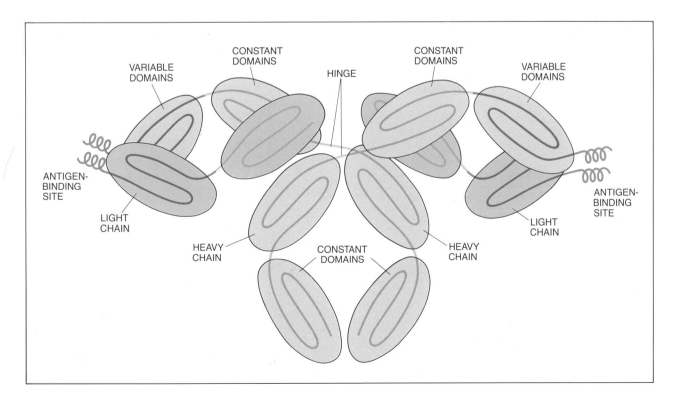

**STRUCTURE OF AN ANTIBODY** can be analyzed in terms of repeated domains, or independent folding units. A light chain consists of two such domains; the heavy chains shown here have four domains, although some heavy chains have three. Within a domain the polypeptide chain assumes a characteristic pattern of folding that includes several strands of the substructure called a beta sheet. The variable region of each polypeptide is confined to a single do- main at the amino end of the chain. Three loops (called hypervari- able regions) within the variable domain contribute to the antigen- binding site. The domain structure shown is schematic; the actual folding pattern is more complex. Similar domains appear in the *T*- cell-receptor protein and in the proteins of the major histocompati- bility complex, which identify an individual's cells. All three fami- lies of molecules have probably evolved from a common ancestor.

tigen include gamma or alpha chains.

For many years there were two competing theories of the genetic origin of antibody diversity. One school of thought held that the germ line (the complement of genes passed from one generation to the next) must include a separate gene for every polypeptide that ultimately appears in an antibody. In this germ-line theory immunoglobulin genes are expressed in exactly the same way as those for any other proteins, and no special gene-processing mechanisms are needed. On the other hand, the model requires an enormous number of immunoglobulin genes.

The second theory supposes there are only a limited number of antibody genes in the germ line, and they somehow diversify as *B* lymphocytes emerge from their stem cells. In other words, the diversification takes place in the somatic, or body, cells rather than in the germ cells.

An interesting variation of the germ-line theory was introduced in 1965 by William J. Dreyer and J. Claude Bennett of the California Institute of Technology. They suggested that for each type of antibody polypeptide chain the germ line includes many *V* genes (one to encode every possible variable region) and a single *C* gene for the constant region. As the cell matures it randomly selects one of the *V* genes and combines it with the *C* gene to create a single length of DNA encoding the full polypeptide.

Dreyer and Bennett's model has certain attractive features. It makes efficient use of the genome, and it offers a natural explanation of how antibody molecules can vary greatly in one part of their structure and remain constant in other parts. Until the mid-1970's, however, there was a major impediment to acceptance of the theory: it required some means of rearranging genes in somatic cells. No such mechanisms were known, and many workers considered them unlikely to exist. The doctrines that one gene always encodes one polypeptide and that the genome remains constant throughout an organism's development were then considered universally established principles of biology.

In the past 10 years the application of recombinant-DNA technology to the study of immunoglobulin genes has shown that they do undergo somatic recombination, but in ways much more complicated than Dreyer and Bennett supposed. Through these complex rearrangements great diversity is generated in the *V* regions.

The first evidence of somatic recombination in immunoglobulin genes was found by Nobumichi Hozumi and me in 1976, when we were both working at the Institute for Immunology in Basel, Switzerland. Our experiments made use of restriction enzymes, which cut DNA at points marked by a particular sequence of nucleotides. The results showed that in embryonic mouse cells the DNA sequences encoding the *V* and the *C* regions of the light chains are some distance apart. In a mature antibody-secreting cell they are much closer together. (The latter finding was based on work not with normal *B* cells but with cells of a myeloma, or lym-

phocyte cancer. Such malignant cells are much easier to grow in culture.)

The mechanisms responsible for the shuffling of immunoglobulin DNA sequences became clearer when fragments of the DNA were cloned in bacteria and analyzed. This was first done by Ora Bernard and me in Basel, in collaboration with Allan Maxam and Walter Gilbert of Harvard University. Working with a DNA clone derived from embryonic mouse cells, we determined the nucleotide sequence of a segment of DNA encompassing the *V* gene of a particular lambda light chain. To our surprise we found that the nucleotides corresponding to the last 13 amino acids of the variable region were missing. They were discovered by Christine Brack of my laboratory. The missing fragment is thousands of base pairs away from the DNA encoding the rest of the *V* region and lies about 1,300 base pairs "upstream" from the start of the *C* region. This small interposed segment has been named *J,* for joining. Each lambda light chain is assembled by combining the scattered *V, J* and *C* segments.

A similar analysis was soon carried out for the kappa light chain and for the heavy-chain variable region. The work was done in several laboratories, notably my own in Basel, Philip Leder's at the National Institutes of Health and Leroy E. Hood's at Caltech. The kappa chain too is encoded by *V, J* and *C* segments. Furthermore, multiple copies of the *V* and *J* segments were discovered: there are a few hundred *V* segments, differing slightly in

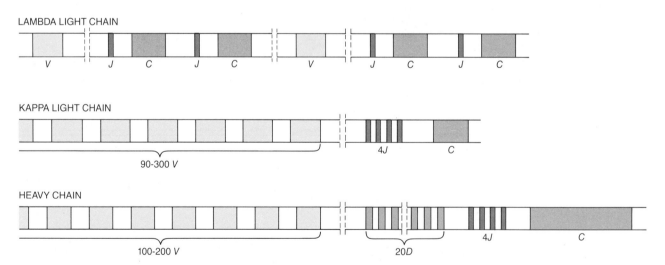

LAMBDA LIGHT CHAIN

V        J      C        J      C            V            J      C        J      C

KAPPA LIGHT CHAIN

90-300 *V*                                            4*J*              *C*

HEAVY CHAIN

100-200 *V*                                20*D*                    4*J*              *C*

**GENES FOR ANTIBODIES** are broken up into small segments scattered widely throughout the genome. Two kinds of light chain appear in mammalian antibody molecules. For the lambda light chain of the mouse there are two *V* genes that encode most of the variable region and four *C* genes for the constant region. Upstream of each *C* gene is a short segment of DNA designated *J,* for joining, which specifies the remainder of the variable region. Either *V* gene can be combined with any pair of *J* and *C* genes. For the kappa light chain there are a few hundred *V* segments, four *J* segments and a single *C* gene. The heavy-chain genes are similar, except that the DNA for the variable region is further subdivided: in addition to the *V* and *J* segments there are about 20 *D* (for diversity) segments. Each set of genes is on a different chromosome. The *T*-cell-receptor genes are organized much as the heavy-chain genes are.

amino acid sequence, and four distinct *J* segments. The number of possible kappa-chain variable regions is the product of these numbers.

The potential diversity of the heavy chains is even greater. In addition to *V* and *J* segments the genes for the heavy-chain variable region include a third fragment designated *D* (for diversity). Mouse germ-line cells have a few hundred *V* segments, about 20 *D* segments and four *J* segments. In principle they can be brought together in well over 10,000 combinations. Combining a light chain with a heavy chain can probably yield more than 10 million distinct antigen-binding sites.

The assembly of a functioning immunoglobulin gene takes place in two stages. First the *V* and *J* segments in a light chain or the *V, D* and *J* segments in a heavy chain are brought together within the DNA. An RNA transcript is then made of the entire length of DNA, including the *V-J* or *V-D-J* complex, the *C* gene and the intron, or noncoding intervening sequence, that separates them. Finally the intron is excised and the messenger RNA is exported from the nucleus to be translated into protein.

The second stage in this process relies on mechanisms of RNA splicing that are common to many families of genes in eukaryotic cells. The first stage, in which the DNA itself rather than the RNA transcript is altered in a highly specific manner, is more unusual and may even be unique to the immune system. It evidently employs a set of enzymes that can bring together distant *V, D* and *J* segments, often deleting all the DNA that separates them. The enzymes themselves have not been isolated, but signal sequences that probably guide their action have been discovered. For example, just downstream of each *V* gene for the kappa chain there is a distinctive pattern composed of a heptamer, or seven-nucleotide unit, followed by a spacer and a nonamer, or nine-nucleotide unit. Just upstream of the *J* segment there is an approximately complementary nonamer-spacer-heptamer pattern. These units could provide a template for the enzymes that cut and rejoin the double helix. Similar signal sequences are found in heavy-chain genes, arranged so that a *D* segment will be included between the *V* and the *J* segments.

The many possible combinations that can be formed from several hundred gene segments are the key to antibody diversity, but there are at least two additional sources of variety. One of these is a lack of precision in the DNA-splicing machinery that fus-

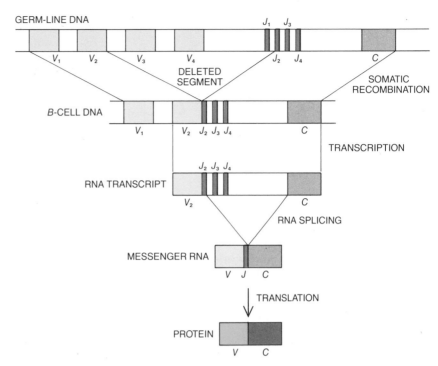

**ASSEMBLY OF AN ANTIBODY GENE** from scattered fragments is done in two stages, shown for a kappa light chain. First randomly selected *V* and *J* segments are fused by enzymes that delete all the DNA lying between them. Here the gene segments labeled $V_3$, $V_4$ and $J_1$ are deleted, bringing together $V_2$ and $J_2$. Next the entire length of DNA from the start of $V_2$ to the end of the *C* gene is transcribed into RNA. Standard RNA-splicing enzymes, which take part in the expression of many genes, excise all the RNA from the end of $J_2$ to the start of *C*. The resulting sequence of messenger RNA is translated into protein.

es *V, D* and *J* segments. The site of the junction can vary by several base pairs. Furthermore, in some cases additional base pairs are inserted in the process of combining segments. Both kinds of change can obviously alter the amino acid sequence of the polypeptide. As a result, even if two antibodies are specified by the same collection of gene segments, they may still have slightly different antigen-binding sites.

Another major source of diversity is somatic mutation. In 1970 Martin Weigert, working in Melvin Cohn's laboratory at the Salk Institute for Biological Studies, determined the amino acid sequences of the lambda light chains derived from 18 mouse myelomas. All the mice were of the same inbred strain and so should have been genetically identical. Weigert found that 12 of the lambda chains were indeed identical but that the other six differed both from the majority sequence and from one another. Spontaneous genetic changes in the developing cells were a likely explanation, but cogent evidence of somatic mutation was not obtained until immunoglobulin genes were cloned and sequenced. In 1977 Brack and Bernard showed that the inbred mouse strain carries only one germ-line *V*-region gene for the lambda chain and that its nucleotide sequence corresponds to the ami-

no acid sequence found in 12 of the myelomas. The logical conclusion is that the six variants arose by somatic mutation.

Since then amino acid sequences have been compared with germ-line nucleotide sequences for a number of kappa light chains and heavy chains. In every case the diversity observed in the proteins has been greater than that of the germ-line DNA. Mutations are seen in the variable domain and in the immediately adjacent regions but not in the constant domains. Estimates of the rate of mutation suggest there should be one change in the *V* region for every three to 30 cell divisions, a rate several orders of magnitude greater than the average mutation rate for eukaryotic genes. It therefore seems that *B* cells or their progenitors carry an enzymatic apparatus for inducing mutations in the variable region of immunoglobulin genes. As yet the nature of the enzymes is entirely unknown.

The presence of both combinatorial and mutational mechanisms for the diversification of antibody genes is intriguing. Why have two systems evolved to meet the same need? Recent studies suggest a plausible explanation. Both mechanisms seem to be under strict control during the development of *B* lymphocytes. The recombination of the immunoglobulin gene

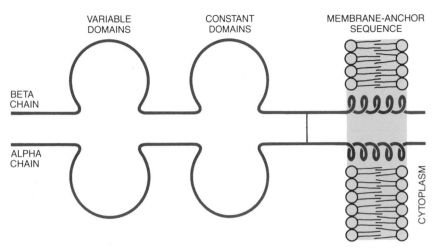

STRUCTURE OF THE *T*-CELL RECEPTOR is not yet known in detail, but its polypeptide components have been identified. Each receptor molecule includes one alpha chain and one beta chain; each chain in turn includes one constant and one variable domain. A region rich in hydrophobic amino acids anchors the protein in the membrane of the *T* cell.

segments is accomplished first, and it is complete by the time the cells are first exposed to antigens. It creates a population of cells that vary widely in their specificity, from which a few cells are selected by any given antigen. The mutational mechanism is then called into action during the proliferation of the selected *B*-cell clones. By altering individual nucleotide bases the mutations fine-tune the immune response, creating immunoglobulin genes ·whose products better match the antigen.

The effects that DNA-joining inaccuracy, base-pair insertion and somatic mutation have on antibody diversity are hard to quantify, but they surely increase the number of distinct antigen-binding sites by a factor of 100, and the factor is probably still larger. Thus if the combinatorial mechanisms alone give rise to 10 million antibodies, the total number could well be greater than a billion.

Given the intricacy of the *B* cells and their antibody-producing machinery, it is somewhat daunting to realize they constitute only half of the immune system. The *T* cells are just as complex and are essential to immunological competence. An animal deprived of *T* cells cannot mount an effective immune response to most antigens even though its *B* cells are intact.

There are three known subpopulations of *T* cells, all of them identical in appearance but distinguished by function. The cytotoxic *T* cells kill their target cells directly. The method of destruction is not known; an activated cytotoxic *T* cell binds to its target but does not engulf it (as a macrophage would), causing a lesion that kills the target cell. The other populations of *T* cells, called *T* helper cells and *T* sup-

pressor cells, have a regulatory role. The *T* helper cells, when they recognize an antigen, stimulate other components of the immune system, including *B* cells and other *T* cells specific for the same antigen. The suppressor cells have just the opposite effect, that is, they diminish the activity of the same groups of cells.

The name *T* helper cell suggests an auxiliary or subordinate role, as if the cells merely abet a response that would take place even in their absence. Actually the *T* helper cells may be the master switch of the immune system. *B* cells, for example, recognize antigens independently of *T*-cell stimulation, but their proliferation and terminal differentiation usually requires activation by *T* helper cells. The *T* suppressor cells would seem to be equally important: by providing negative feedback they make the immune response self-limiting. They may also have a part in eliminating *B* and *T* cells directed against the self.

Because the *T* cells are antigen-specific, they must have receptor molecules analogous to the membrane-bound immunoglobulins of *B* cells. This fact was recognized more than 20 years ago, but the *T*-cell receptors proved difficult to analyze or even identify because they are not secreted in large quantities the way antibodies are. The receptors were first glimpsed in experiments done by James P. Allison of the University of Texas at Austin, John W. Kappler of the National Jewish Hospital in Denver and Ellis L. Reinherz of the Harvard Medical School. They prepared antibodies that bind to a protein on the *T*-cell surface; the protein identified in this way was considered a good candidate for the role of a receptor because it varies in

structure from one clone of cells to another. The mass of the protein is about two-thirds that of an immunoglobulin, and it consists of two subunits, designated alpha and beta.

In 1984 Tak W. Mak and his colleagues at the University of Toronto and Mark M. Davis and his colleagues at the Stanford University School of Medicine cloned and sequenced a gene that is expressed and rearranged in *T* cells but not in *B* cells. Mak worked with cells from a human *T*-cell leukemia and Davis with a hybridoma, a cell line created by fusing a mouse *T* helper cell with a malignant *T* cell. In spite of the disparate origins of the two genes they were found to encode the same protein.

The nucleotide sequences analyzed by Mak and Davis are homologous to those of immunoglobulin genes, and there are also larger-scale indications of a familial resemblance to immunoglobulins. The genes are divided into scattered segments that can be rearranged in the developing *T* cell, and the upstream segments (corresponding to the amino end of the polypeptide) are variable whereas the downstream segments have a constant sequence. As in membrane-bound immunoglobulins, a series of hydrophobic amino acids near the carboxyl end of the protein anchor the molecule in the membrane. A direct determination of amino acid sequences by Reinherz and his coworkers has since confirmed that the genes specify the beta subunit of the *T*-cell receptor.

Two more *T*-cell DNA clones were isolated by Haruo Saito, working in my laboratory at the Massachusetts Institute of Technology, and David M. Kranz in Herman N. Eisen's laboratory, also at M.I.T. In this case the genes came from mouse cytotoxic *T* cells; nevertheless, the downstream part of one clone is essentially identical with the constant-region gene for the helper-cell beta chain. The second DNA sequence has a number of properties in common with the beta-chain genes. It is homologous to immunoglobulin genes, is made up of segments that are rearranged and expressed only in *T* cells and has a hydrophobic anchor segment. The logical hypothesis was that the gene encodes the alpha chain of the receptor molecule.

Soon afterward, however, Saito isolated a third gene from the same clone of cytotoxic *T* cells. It too has all the properties expected of a *T*-cell receptor, and it has an additional factor in its favor. Chemical analysis of the protein, done in parallel with the gene-cloning studies, showed that the receptor protein has carbohydrate side chains attached to it through the amino acid asparagine. The earlier alpha-chain candidate has no asparagine units in the appropriate positions, whereas the new one has several suitable asparagines. A partial determination of the amino acid sequence of the alpha subunit by Kappler and his co-workers has confirmed that the third of the DNA clones is the true alpha-chain gene. The same gene was also isolated from a helper-cell hybridoma by Y.-H. Chien and others in Davis' laboratory at Stanford.

In this account the second gene found by Saito and Kranz—the discarded alpha-chain candidate—would seem to be left without a function. It is so closely related to the other genes, however, that it almost certainly has some role in the recognition of antigens. The protein it is presumed to encode is now designated the gamma chain; I shall discuss below what part it might play in the T cell's action.

From the nucleotide sequences specifying the alpha and beta chains much of the structure of the T-cell receptor can be inferred. Each chain is composed of two domains, which are similar in overall structure to the repeated domain of the immunoglobulins. The degree of sequence homology with the immunoglobulins is between 25 and 35 percent. The two chains are linked by a disulfide bond between the constant region and the membrane-anchoring peptide. Molecules derived from helper cells and cytotoxic T cells have identical constant regions in both the alpha and the beta chains.

The molecular genetics of the T-cell-receptor molecules is also strikingly similar to that of immunoglobulins. The variable regions of both receptor chains are encoded by three gene segments, corresponding to the V, D and J segments, which are scattered along a chromosome in germ-line cells but are fused in mature T lymphocytes. The heptamer-nonamer signal sequences associated with immunoglobulin genes are also found near the T-cell-receptor segments, indicating that the same system of enzymes or a very similar one is employed to mediate the somatic recombination.

Considering all the similarities between genes for immunoglobulins and those for T-cell receptors, it seems reasonable to speculate that the two kinds of protein might recognize antigens in the same way. In other words, the T-cell receptors might have an antigen-binding site formed by clusters of highly variable amino acids in specific subregions within the variable regions of the alpha and beta chains. It is an appealing hypothesis in that it supplies a single explanation for the recognition abilities of both proteins. Even if the hypothesis turns out to be correct, however, it cannot be the whole story. The reason is that the two branches of the immune system recognize antigens in different circumstances. A B cell can respond to an antigen alone, but an individual's T cells are activated only if the antigen is displayed on the surface of a cell that also carries markers of the individual's own identity.

To describe this difference in antigen response it is necessary to introduce the molecules that serve as markers of individual identity. They are proteins encoded by a large cluster of genes called the major histocompatibility complex, or MHC, and they make up a third class of proteins with a vital role in immune recognition.

The MHC proteins were discovered in tissue-grafting experiments. Unless the donor and the recipient of a graft are genetically identical (as in the case of identical twins or mice of an inbred strain) the graft is generally rejected because the recipient mounts an immune response to the donor's MHC proteins. The prevalence of graft rejection implies that unrelated individuals almost always express different sets of MHC genes. Indeed, apart from immunoglobulins and T-cell receptors, the MHC proteins are the most diverse ones known. Whereas antibodies and T-cell receptors vary from one cell to the next, however, the MHC proteins

SINGLE-RECEPTOR MODEL

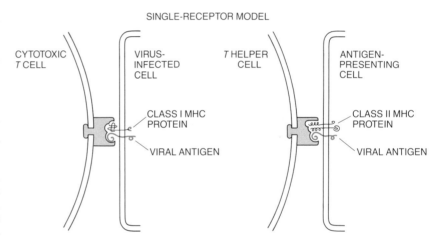

DUAL-RECEPTOR MODEL

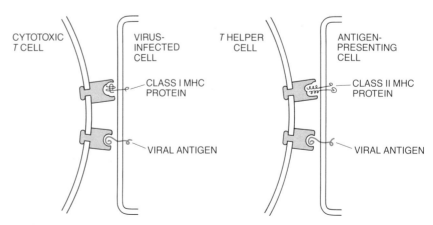

T-CELL-RECEPTOR SYSTEM, unlike an antibody molecule, does not respond to an antigen alone; the antigen must be presented on the surface of a cell that also displays one of the proteins of the major histocompatibility complex (MHC). Cytotoxic T cells recognize an antigen in combination with a Class I MHC protein, found on almost all body cells. T helper cells bind to an antigen associated with a Class II MHC protein; the two molecules are confined to cells of the immune system, such as macrophages and lymphocytes. It is not yet clear whether T cells have a receptor with two binding sites or two separate receptor molecules.

differ from one individual to another.

Two classes of MHC proteins have been identified. A Class I molecule consists of a large polypeptide chain (about the size of an immunoglobulin heavy chain) linked to a much smaller subunit called beta-2 microglobulin. Class I MHC proteins are found on the surface of virtually all cells. The Class II proteins, in contrast, appear only on a few types of cells that have a part in the immune response, such as *B* lymphocytes, macrophages and specialized epithelial cells. A Class II molecule is also made up of two polypeptide chains, both about the size of an immunoglobulin light chain. All the MHC polypeptides exhibit some degree of homology with immunoglobulins, although the resemblance is not as strong as that between *T*-cell receptors and immunoglobulins.

Since the transfer of tissue between individuals is rare in nature, graft rejection cannot be the primary function of the MHC proteins. Their true purpose lies elsewhere in the immune system, namely in directing the responses

of *T* cells. A *T* cell recognizes both an antigen and a self MHC protein on the surface of a single cell. The requirement of dual stimuli is called MHC restriction. Cytotoxic *T* cells respond to antigen together with a Class I MHC protein; *T* helper cells require a Class II protein.

Of what benefit to the organism is MHC restriction? Its effect is to direct the activities of *T* cells to the surface of the animal's own cells rather than to bacteria or free foreign molecules. One plausible idea is that the cytotoxic *T* cell arose to provide protection against viral infection. When a virus enters a cell, coat proteins encoded by the viral genome are displayed on the cell membrane. Hence the infected cell has just the right pattern of surface markers for recognition by *T* cells: a foreign molecule in combination with native proteins. A cytotoxic *T* cell, recognizing the viral antigen and one of the Class I proteins, can kill the infected cell before the virus replicates.

The Class II proteins and the regulatory *T* cells may have evolved to in-

crease the efficiency of the immune response. *T* helper cells can be triggered by antigen-presenting cells that take up circulating antigens and display them on their surface along with Class II MHC proteins. The presence of Class II molecules on *B* lymphocytes and macrophages may be the key to how helper cells communicate with those cells and thereby recruit their participation in an immune response.

If a *T* cell must recognize two surface markers, the question arises of whether it has two separate receptors or a single dual-function receptor. Some recent experiments seem to favor the single-receptor model, but the results are by no means conclusive. If a second receptor is eventually found, it may turn out to incorporate the "orphaned" gamma chain, which has all the properties expected of a receptor protein but has no place in the current scheme of *T*-cell operations.

There is another possible role for the gamma chain. *T* lymphocytes become mature and functional only after a pe-

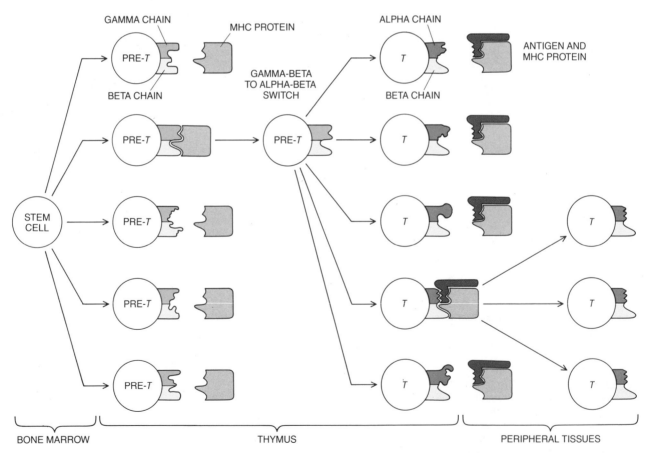

**THYMIC "EDUCATION"** is a necessary stage in the development of functional *T* lymphocytes. A model proposed by David Raulet and the author offers one possible explanation of the developmental process. According to the model, immature *T* cells first make receptor proteins with polypeptide chains called gamma (*color*) and beta (*light color*). In the thymus the lymphocytes are exposed to MHC proteins, and only those cells with sufficient affinity for these markers of self-identity are allowed to propagate. If the selected cells were released from the thymus, however, they would attack the body's own tissues. The affinity of the receptor molecules for self-antigens must therefore be reduced. Each receptor retains the beta chain of the selected clone, but the gamma chain is replaced by one of a variety of alpha chains (*solid color*). The modified *T* cells respond to a self-MHC protein in combination with an antigen.

riod of residence in the thymus, and it is during this period of "thymic education" that *T* cells come to recognize antigens only in combination with the individual's own MHC proteins. No definitive explanation of thymic education has been given, but many immunologists agree that the crucial step must be selection of a subpopulation of immature *T* cells through their interaction with self MHC proteins displayed to them by thymic cells. According to one model, each immature *T* cell responds to just one MHC protein (or to a small group of them), but the total population includes cells responsive to all possible markers. In the thymus only those cells that have sufficient affinity for the native MHC proteins are allowed to propagate and continue their differentiation.

For this scheme to work immature *T* cells must be able to recognize and respond to MHC molecules alone, without an accompanying antigen. When the mature *T* cells are released from the thymus, they have obviously lost that capability; otherwise they would be aggressive against the body's own tissues. What is the biochemical basis of the change in reactivity? Recent studies indicate that in immature *T* cells the alpha gene is expressed at low levels, whereas the beta and gamma genes produce larger quantities of protein. On the basis of these findings David Raulet of M.I.T. and I have proposed a model of *T*-cell development that we call the gamma-beta-to-alpha-beta switch.

In the model immature cells have receptors made up of a gamma chain and a beta chain and are responsive to MHC proteins alone. In the course of differentiation the gamma gene is turned off and the alpha gene is turned on, so that mature cells have alpha-beta receptors. The change reduces the cell's affinity for self MHC proteins but, because the beta chain is still present, does not extinguish it entirely. An analogous mechanism operates in red blood cells when they switch from the fetal to the adult form of hemoglobin.

Our proposal for the gamma chain's function has not yet been tested, but the tools are now in hand to settle this question and many others about the nature of the *T*-cell receptor. Ideally, structural and genetic studies would yield an understanding of these molecules at the same level of detail that can now be given for the immunoglobulins. At that point one might hope to resolve some of the major remaining enigmas of immunology: how *T* lymphocytes develop in the thymus, how they recognize their target cells and how they control the rest of the immune system.

# 8

# THE MOLECULAR BASIS OF COMMUNICATION BETWEEN CELLS

# The Molecular Basis of Communication between Cells

*Chemical messengers mediate long-range hormonal communication and short-range communication between nerve cells. The two systems differ in directness, but some messenger molecules are common to both*

by Solomon H. Snyder

The unicellular amoeba can perform every function necessary to sustain life. The cell can assimilate nutrients from the environment, move itself about and carry out the metabolic reactions that provide it with energy and synthesize new cellular molecules. In multicellular organisms the situation is considerably more complex. The various tasks are split up among many distinct cell populations, tissues and organs, which may be far apart from one another. To coordinate all these various functions there must be mechanisms whereby individual cells or groups of cells can communicate with one another.

In most higher organisms there are two primary methods of communication between cells: systems of hormones and systems of neurons, or nerve cells. In both systems cells "talk" to one another by means of chemical messengers. The main difference between the two systems is the level of directness with which they act. A neuron sends discrete messages to a specific set of target cells: muscle cells, gland cells or other neurons. To send a message the neuron releases a chemical called a neurotransmitter toward the target cell. The cell-to-cell communication takes place at specific sites known as synapses. Molecules of neurotransmitter become attached to receptors (usually protein molecules) on the surface of the target cell and effect chemical changes at the cell membrane and within the cell itself.

Hormone action is usually less direct. Although so-called autocrine and paracrine mechanisms exist, in which a hormone acts respectively on the cell that released it or on an adjacent cell, the commonest form of hormonal communication is in the endocrine system, where a gland releases hormones that may act on cells or organs any-where in the body. Endocrine glands secrete hormones into the bloodstream; each target cell is equipped with receptors that recognize only the hormone molecules meant to act on that cell. The receptors pull the hormone molecules out of the bloodstream and bring them into the cell.

There are thus considerable differences between hormonal and neural communication. Neurons tend to act over short distances on a particular cell or set of cells; neuronal communication can take place in a few milliseconds. In contrast, a hormone released by a particular gland may go on to affect cells or organs in virtually any part of the body; hormonal communication can take several hours.

Yet on the molecular level these two systems are not as dissimilar as they at first seem. Both operate by causing special messenger molecules to come in contact with specific receptors on the target cell. In addition certain neurotransmitters seem, like hormones, to serve only in specialized systems of communication and to perform specific functions.

Recently it has become apparent that there is an even closer relation between the two major systems of intercellular communication: many of the messenger molecules employed by one system are also employed by the other. For instance, norepinephrine, as a hor-mone, is released by the adrenal gland to stimulate contractions of the heart, dilate the bronchial tract of the lungs and increase the contractile strength of arm and leg muscles. On the other hand, norepinephrine is also a neurotransmitter: in the sympathetic nervous system it constricts blood vessels, thereby increasing blood pressure.

The same kind of messenger molecule can carry a very different message in the hormonal system than it does in the nervous system. It must be that certain molecules are particularly good mediums of communication.

Molecules that act as hormones generally fall into one of two chemical classes: peptides and steroids. The peptides are strings of amino acids (the subunits of proteins). The steroid hormones are large molecules derived from cholesterol that share the same basic structure: 17 carbon atoms bound together in four closely linked rings [*see illustration on page 88*]. Small differences in the chemical groups attached to the carbon rings give rise to hormones with quite different functions. Among the major steroid hormones in human beings are the glucocorticoids called cortisol and corticosterone, which regulate the metabolism of glucose and control a wide range of other metabolic functions; the mineral steroids, such as aldoster-

**VASOPRESSIN MOLECULE carries messages in both the hormonal and the neuronal systems of communication between cells. As a hormone, vasopressin is released by cells in the posterior pituitary gland. It raises blood pressure by constricting the blood vessels, and it acts as an antidiuretic by increasing the kidneys' ability to reabsorb water. Vasopressin is also a neurotransmitter: a substance that carries messages from one nerve cell to another. As a neurotransmitter, vasopressin is found in the brain, where it is thought to have a role in the mechanisms of memory. In this image, made by Tripos Associates of St. Louis, solid lines represent bonds between atoms; dotted surfaces delineate the surface of the molecule. The color of the bonds and dots shows which atoms occupy a region: white is for carbon or hydrogen, blue for nitrogen, red for oxygen, orange for phosphorus and yellow for sulfur.**

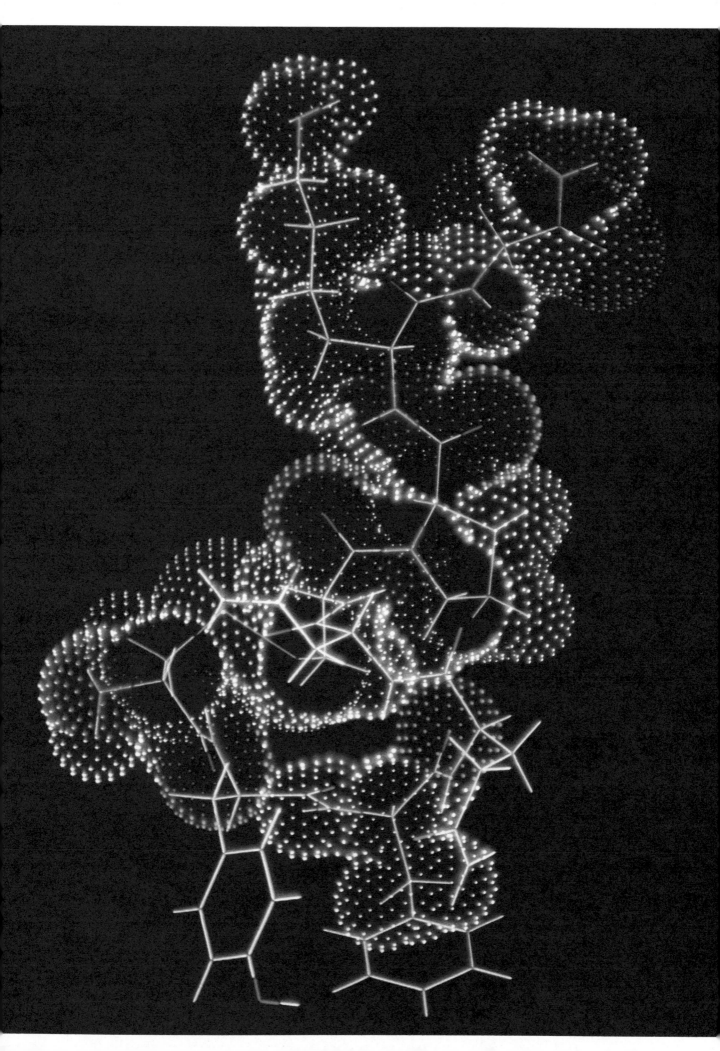

one, which affect the body's salt balance, and the sex steroids, which include progesterone, testosterone and the estrogen hormones.

The female sex steroids (the estrogens and progesterone) have been characterized more completely than most hormones in the body. They provide an excellent example of how hormones are produced, released and regulated. Estradiol, which is the major estrogen hormone, and progesterone combine during the normal menstrual cycle to prepare the uterus for implantation of a fertilized ovum by building up the uterine wall and increasing blood flow to the uterus. It is an abrupt decline in estradiol and progesterone levels that triggers the bleeding associated with menstruation.

The release of the sex steroids, like that of most hormones, is itself controlled by other hormones and so-called releasing factors from the two master organs: the pituitary and the hypothalamus. Generally speaking, the hypothalamus initiates the release of a steroid hormone from a peripheral gland by releasing a factor that acts on the pituitary. The pituitary then releases other hormones, which act on the peripheral glands. The peripheral glands respond by releasing hormones that act on the target cells.

In the case of the female sex steroids, the major factor released by the hypothalamus is called gonadotropin-releasing hormone. At the beginning of the monthly menstrual cycle the hypothalamus, influenced by regions of the brain that act as timers, secretes gonadotropin-releasing hormone (which is sometimes called luteinizing-hormone releasing hormone) into the bloodstream. The releasing hormone acts at receptors on the surface of cells within the pituitary that release hormones known as luteinizing hormone and follicle-stimulating hormone.

The follicle-stimulating hormone and to a lesser extent the luteinizing hormone stimulate the growth of small, round follicles in the ovary. The follicles convert cholesterol into estradiol, which they release into the bloodstream. Estradiol goes on to build up muscle tissue in the wall of the uterus. Several days later the pituitary releases a quantity of luteinizing hormone, which changes the structure of the ovarian follicles to form a new entity called the corpus luteum. The transformed follicles synthesize less estradiol and begin to convert cholesterol into progesterone. Progesterone increases blood flow to the uterus and slows the uterine contractions. The combination of estradiol and progesterone has thus prepared the uterine wall to receive a fertilized egg. Soon after ovulation the pituitary begins to release less luteinizing hormone, causing the corpus luteum to stop synthesizing progesterone. Then the cells lining the uterus are sloughed off and menstrual bleeding starts.

Throughout the menstrual cycle the amounts of hormones secreted by various glands must be controlled to ensure that the bloodstream contains the proper concentration of each hormone. This control is achieved through an elaborate set of feedback mecha-

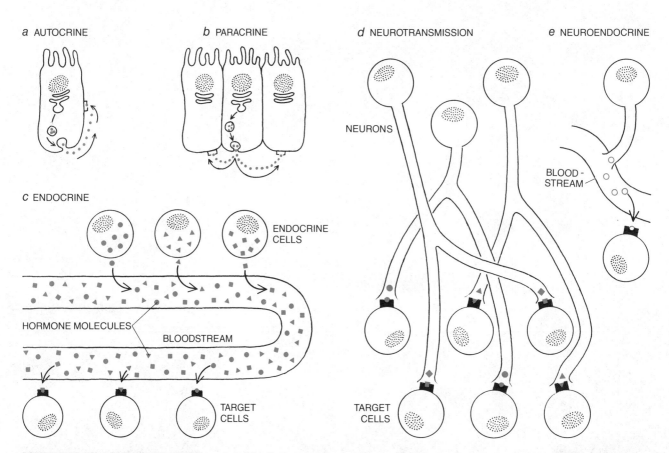

**METHODS OF COMMUNICATION** employed by the hormonal system are generally less direct than those employed by the nervous system. Although autocrine hormones (*a*) act on the cell that releases them and paracrine hormones (*b*) act on adjacent cells, most hormones are in the endocrine system and act on cells or organs anywhere in the body. Endocrine glands (*c*) release hormone molecules into the bloodstream, where they come in contact with specific receptors on target cells. A cell's receptors recognize the hormones meant to act on that cell and pull them out of the bloodstream. Neurons (*d*) communicate by releasing neurotransmitters close to specific target cells. Neural communication is characterized by discrete messages sent over short distances. Some neurons, however, have a role in the hormonal system: in neuroendocrine action (*e*) a neuron releases substances that act as hormones directly into the blood.

nisms [*see illustration on page 89*]. For example, estradiol released in the ovaries acts not only on its target cells within the uterus but also on the cells in the pituitary that release follicle-stimulating hormone. There it prevents the pituitary from inducing the ovaries to produce more estradiol. Estradiol also acts on the hypothalamus, where it inhibits the release of gonadotropin-releasing hormone. The amount of estradiol in the bloodstream thus determines the additional amount to be released, just as the amount of heat inside a house determines, by way of the thermostat, how much more heat the furnace is to generate.

In the case of estradiol there is another feedback loop as well. When the follicles in the ovaries produce estradiol, they also produce a substance called inhibin. Inhibin acts on both the pituitary and the hypothalamus: in the pituitary it restricts production of follicle-stimulating hormone, and in the hypothalamus it restricts production of gonadotropin-releasing hormone.

Other steroid hormones, which have widely varying functions, follow similar principles of feedback and control. For instance, the glucocorticoids, such as cortisol, are formed and released when factors and hormones from the hypothalamus and pituitary stimulate the adrenal cortex. Whereas estradiol acts on only a few specific target organs, cortisol influences nearly every tissue in the body, causing metabolic changes that increase the organism's ability to deal with continuous stress. In most tissues cortisol enhances the uptake and conversion into protein of amino acids, and in the liver it accelerates the conversion of amino acids into sugars. The adrenal cortex is induced to form and secrete cortisol by corticotropin, a hormone synthesized in the pituitary. As Wylie Vale of the Salk Institute for Biological Studies has recently shown, the secretion of corticotropin is in turn controlled by the hypothalamus, which secretes a so-called corticotropin-releasing factor.

The hypothalamic releasing factors that regulate the pituitary and the pituitary "master" hormones, which in turn regulate the release of steroid hormones, are themselves not steroids. They are members of the other major chemical class of hormones, the peptide hormones.

Unlike the steroid hormones, which are all synthesized from the same molecule (cholesterol), each peptide hormone derives from a specific "precursor" molecule: a long string of amino acids that contains one or more copies of the hormone as well as other, unre-

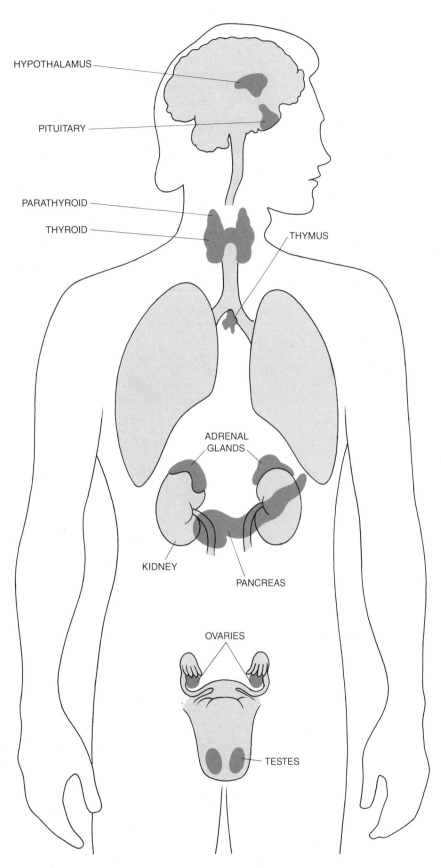

**ENDOCRINE SYSTEM consists of several distinct glands and control centers. Endocrine glands are all controlled by the pituitary, in a sense the master gland: it secretes hormones that stimulate other glands to synthesize and release their own hormones. The pituitary is in turn controlled by the hypothalamus, which is not a gland but a distinct region of the brain; releasing factors secreted by the hypothalamus control the release of pituitary hormones.**

lated peptide sequences. The precursor molecule, which is also called a prohormone, is cleaved by enzymes to release the peptide hormone. Although each peptide hormone is derived from a different precursor, some of the processing enzymes are shared by several systems of peptide hormones.

One of the major peptide hormones is insulin, which is released by specific cells, called the beta cells, in the pancreas. Insulin affects nearly every cell in the body. Although it is best known for lowering the concentration of sugar in the blood by increasing the ability of cells to take up glucose, insulin has many other functions as well, some of which are understood only vaguely. For instance, insulin somehow influences fat metabolism in a way that lowers the concentration of fatty substances circulating in the blood. Hence diabetics, who suffer from insulin deficiency, often develop atherosclerosis, in which plaques of fatty deposits line the blood vessels.

Many of the peptide hormones act in the intestine. Gastrin, for example, is a peptide hormone that stimulates the secretion of acid in the stomach. Patients with gastrin-producing tumors often develop severe ulcers owing to excessive amounts of acid production. Somatostatin is a peptide hormone that counters the effects of gastrin by blocking the secretion of acid at discrete cell groups within the stomach. Another gut peptide, cholesystokinin, is released into the bloodstream by certain cells in the intestine; it travels to the gall bladder, where it increases the flow of bile into the intestine, enhancing digestion.

Cholesystokinin has another function as well: it acts as a neurotransmitter in the brain. Many of the other peptide hormones exhibit the same kind of dual capability, serving as messenger molecules in both hormonal and neural communication. For instance, vasoactive intestinal polypeptide is an intestinal hormone that regulates the motility of the gut, but it is also a brain neurotransmitter. The enkephalins, two peptides that differ slightly from each other, act as opiates in the brain, but in the intestine they are hormones that regulate the movement of food through the digestive pathway by altering the rhythmic peristaltic contractions.

Until the mid-1970's it was not known that peptides could act as neurotransmitters, although many of the peptide hormones were well known. The first neurotransmitters to be identified were acetylcholine, the amines (which are slightly modified chains of amino acids) and the monoamines (modified single amino acids). Examples of monoamines include the catecholamines dopamine, norepinephrine and epinephrine. These three molecules are all derived from the amino acid tyrosine. Norepinephrine and epinephrine (which are also known as noradrenaline and adrenaline) are hormones as well as neurotransmitters; they are released from the adrenal medulla, and they act to increase the heart rate and blood pressure and to increase the flow of sugar into the bloodstream from the liver.

In the 1960's many groups of investigators obtained evidence that a number of unmodified amino acids sometimes serve as neurotransmitters. One such amino acid, gamma-aminobutyric acid, functions almost exclusively as a neurotransmitter. Other amino acid neurotransmitters, such as glutamic acid, aspartic acid and glycine, are constituents of proteins as well.

In all there are no more than 10 neurotransmitters that are amines or amino acids. According to the classical model of neurotransmitter function, any more would be unnecessary. In this model a neurotransmitter acts as a chemical switch that either causes its target neuron to fire or inhibits it from firing. If the model were accurate, the brain would be able to function with

STEROID HORMONES, all derived from cholesterol, share a common structure of 17 carbon atoms bound in four closely linked rings. Differences in the chemical groups attached to the carbon rings give rise to hormones with very different functions. The molecules shown represent the primary steroid hormones. Cortisol and corticosterone promote formation of glucose in the liver and therefore are called the glucocorticoids. Aldosterone enables the kidney to retain sodium rather than excreting it in the urine. Testosterone is the principal male sex hormone, and progesterone and estradiol are the major female sex hormones.

only two neurotransmitters, one excitatory and the other inhibitory. Yet since 1975, when it was discovered that the enkephalins are neurotransmitters that act at opiate receptors in the brain, investigators have isolated almost 50 additional "neuropeptides." What do they do?

Careful electrophysiological studies have shown that different neurotransmitters can produce many different effects at synapses. There are a number of different kinds of pores, or channels, in the cell membrane of the target neuron. These channels, which can be opened or closed by a neurotransmitter molecule, allow such ions as chloride, sodium, potassium and calcium to pass through the neuronal membrane. There appear to be many channel types for each ion, and various channels convey different types of electrical information. Neurotransmitters can affect the channels in different ways.

Moreover, a single neurotransmitter may have differing effects depending on the type of synapse at which it is acting. For example, at muscarinic receptors, which are found on smooth-muscle cells in the intestine, the neurotransmitter acetylcholine closes certain channels that normally allow potassium ions to leave the cell. The effect (which is not completely understood) is to produce a gradual, prolonged excitation of the muscle. On the other hand, at nicotinic receptors, which are found on skeletal muscles, acetylcholine opens sodium channels, causing the muscle to contract briskly and rapidly.

In addition to overturning the conception that neurotransmitters deliver only a simple "on" or "off" message, neuropeptides have caused another aspect of traditional thinking to be reconsidered. It had been thought that each neuron releases only a single type of neurotransmitter. In 1977 Tomas G. M. Hökfelt of the Karolinska Institute in Stockholm found that the terminals of many (and perhaps most) neurons contain two or three neurotransmitters. (One of the neurotransmitters is invariably a peptide.) Apparently the "co-transmitters" work together in a synergistic fashion to convey subtler information than would be possible with a single transmitter. It is not yet clear precisely how the mechanisms of co-transmission operate.

Many features of neuropeptides are exemplified by the two enkephalins. Both of them are five amino acids long. They differ only in their fifth amino acid: one, called met-enkephalin, ends with methionine and the other, leu-enkephalin, with leucine. Enkeph-

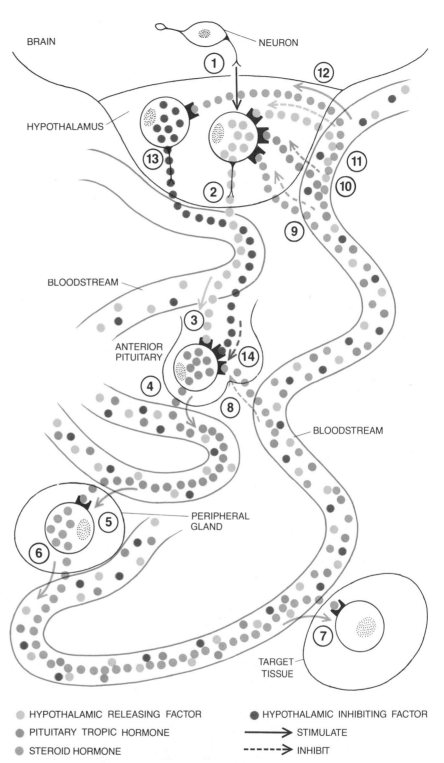

HYPOTHALAMIC RELEASING FACTOR    HYPOTHALAMIC INHIBITING FACTOR
PITUITARY TROPIC HORMONE    STIMULATE
STEROID HORMONE    INHIBIT

**FEEDBACK AND CONTROL MECHANISMS** initiate the release of steroid hormones and ensure that appropriate amounts of a hormone are in the blood. When the hypothalamus is stimulated by neurons in the brain (*1*), it secretes a hypothalamic releasing factor (*2*) into the bloodstream. Some molecules of the releasing factor are pulled out of the blood by specialized receptors on the surface of certain cells in the pituitary (*3*) and induce the cells to produce and secrete a pituitary tropic hormone (*4*). The pituitary tropic hormone travels through the blood to a peripheral gland (*5*) and causes the gland to start producing, say, a steroid hormone (*6*). The steroid hormone goes on to affect its target tissue (*7*). Several feedback loops maintain the correct concentrations of steroid hormone in the bloodstream. For example, the steroid hormone itself acts on the pituitary (*8*) to inhibit the production of pituitary tropic hormone; it also acts on the hypothalamus to limit the production of hypothalamic releasing factor (*9*). The pituitary tropic hormone and the hypothalamic releasing factor itself also inhibit the hypothalamus from producing the releasing factor (*10, 11*). In addition the steroid hormone (*12*) induces certain cells in the hypothalamus to produce a hypothalamic inhibiting factor (*13*), which acts to inhibit the release of the pituitary tropic hormone (*14*).

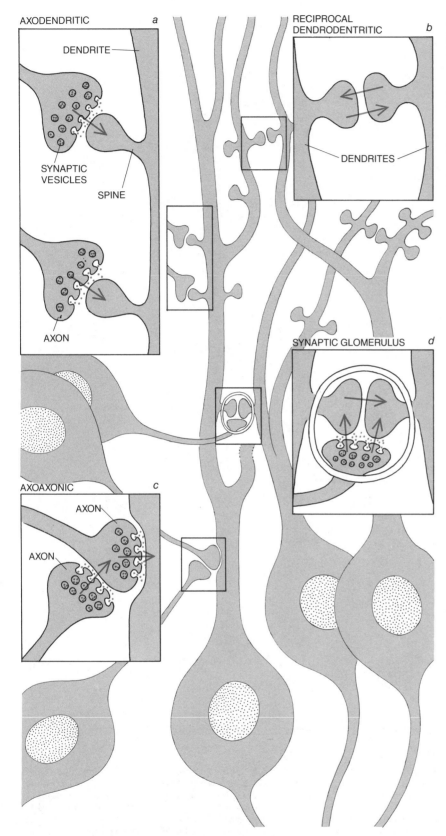

**COMMUNICATION BETWEEN NEURONS** takes place across gaps called synapses. In the classical axodendritic synapse (*a*) synaptic vesicles in the axon of one neuron release neurotransmitter toward receptors on the dendrite of a target neuron. It is also possible for a dendrite to pass a message to another dendrite; such messages are passed by way of dendrodendritic synapses. In a reciprocal dendrodendritic synapse (*b*) each dendrite passes messages to the other by way of a separate synapse. In some other synapses, called axoaxonic synapses (*c*), the axon of one neuron passes a message through the axon of another neuron to the dendrite of a third neuron. In a synaptic glomerulus (*d*) the axon of one neuron passes messages to dendrites of two others; the dendrites may pass messages to each other as well.

alins are observed in discrete neural pathways throughout the brain in approximately the same areas where opiate receptors are found. The coincidence of the location of enkephalin pathways and opiate receptors supports the hypothesis that the enkephalin neurotransmitters act as opiates.

The enkephalins are synthesized from two large precursor proteins, which are called proenkephalin *A* and proenkephalin *B*. Proenkephalin *A* contains six copies of met-enkephalin and one of leu-enkephalin, whereas proenkephalin *B* contains three copies of leu-enkephalin and none of met-enkephalin. Hence all met-enkephalin derives from proenkephalin *A;* leu-enkephalin can be synthesized from either of the two precursors.

Within the precursor molecules each enkephalin is flanked on both sides by two-amino-acid signals: either two lysines or two arginines or one of each. Two successive enzymatic steps are necessary to free enkephalin from the precursor. In the first step an enzyme cuts the peptide bond to the right of each of the flanking amino acids [*see illustration on opposite page*]. The resulting molecule consists of an enkephalin with one extra amino acid attached to the right-hand side. A second enzyme then removes this amino acid to produce enkephalin.

The rate at which enkephalin is formed varies under different circumstances; the rate is thought to accelerate during periods of painful stress. The rate of synthesis can be governed by changes in the rate at which the genes for proenkephalin are transcribed or by the availability of the enzymes that cleave the peptide bonds holding enkephalin within the precursor molecule.

Many enzymes can carry out the two enzymatic steps necessary to convert proenkephalin into enkephalin. One of the most persistent puzzles raised by the neuropeptides has been whether certain generalized enzymes process all hormone and neurotransmitter peptides or whether there are specific enzymes for each peptide. The solution to the puzzle will have profound practical consequences. If specific enzymes are involved, it may be possible to design therapeutic drugs made up of selective enzyme inhibitors that block the biosynthesis of only certain peptide neurotransmitters, allowing precise control over a patient's neurochemistry.

Recent research favors the hypothesis that specialized converting enzymes do indeed exist for each neuropeptide. Lloyd D. Fricker, Stephen M. Strittmatter and David R. Lynch in my laboratory at the Johns Hopkins University School of Medicine have isolated and characterized an enzyme we call enkephalin convertase. It removes the single amino acid that is attached to partially liberated enkephalin. We have found that enkephalin convertase is selectively localized in the same sites in the brain as enkephalin itself, indicating that it contributes to formation of enkephalin in these areas (although it may have other functions elsewhere in the body). Several drugs we have tested are about 1,000 times more potent at inhibiting enkephalin convertase than any other enzyme.

One of the most exciting consequences of research on neuropeptides, then, is the possibility of developing new drugs that are more effective, more specific and safer than the

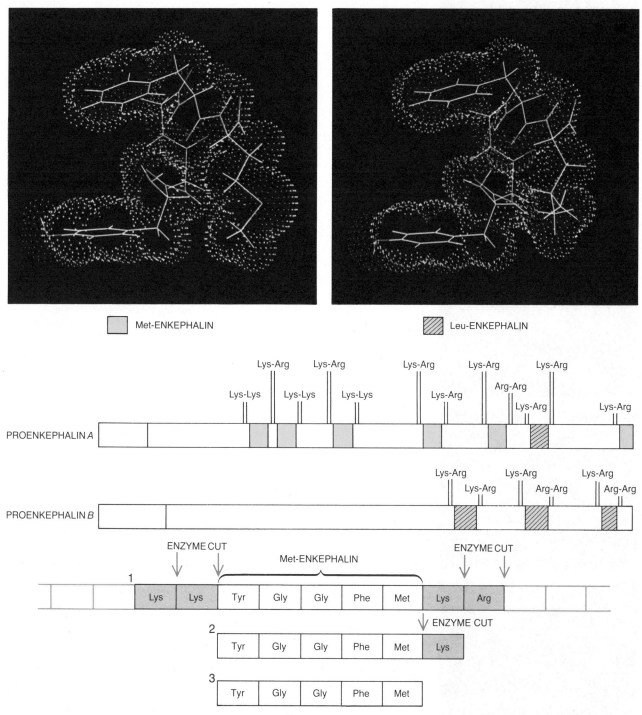

ENKEPHALINS, molecules that act both as neurotransmitters and as hormones, are formed when enzymes cleave much larger "precursor" molecules. Each enkephalin is a string of five amino acids; the two enkephalins differ only in the fifth amino acid, which is methionine in met-enkephalin (*top left*) and leucine in leu-enkephalin (*top right*). There are two types of enkephalin precursor molecule (*middle*). Proenkephalin *A* contains six copies of met-enkephalin and one of leu-enkephalin; proenkephalin *B* contains three copies of leu-enkephalin and none of met-enkephalin. Within the precursor molecules each copy of enkephalin is flanked on both sides by pairs of amino acids that serve as signals; a signal can be composed of two lysines, two arginines or a lysine and an arginine. Two enzymes are needed to separate enkephalin from the precursor molecule. The first (*1*) makes a cut to the right of each signaling amino acid (lysine or arginine), leaving an enkephalin molecule with one extra amino acid attached to its right-hand side. The second enzymatic step (*2*) cuts off the extra amino acid to produce enkephalin (*3*). The images at the top were made by Tripos Associates.

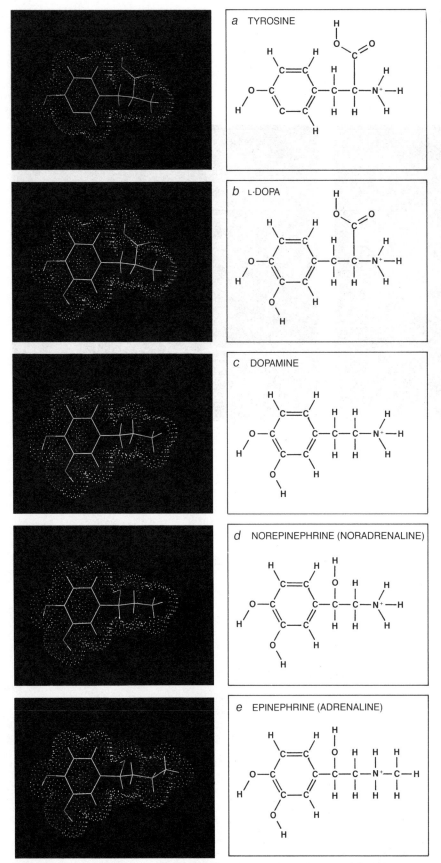

**CATECHOLAMINE NEUROTRANSMITTERS,** dopamine, norepinephrine and epinephrine (which also act as hormones), derive from the amino acid tyrosine. Tyrosine (*a*) is first converted into L-dopa (*b*) by the addition of a hydroxyl (OH) group. Hydrogen replaces a carbon bonded to an OH group and an oxygen to form dopamine (*c*). Adding another OH group yields norepinephrine (*d*), or noradrenaline. Adding a methyl group (CH₃) gives epinephrine, or adrenaline (*e*). The images at the left were made by Tripos Associates.

psychotherapeutic drugs now in use. Virtually every drug used in psychiatry and neurology acts by blocking or enhancing the effects of one or another neurotransmitter. Most drugs exert their effects by acting on one or another of the classical neurotransmitters. The newly discovered peptide transmitters may spawn a new generation of drugs that influence the synthesis, release and receptor effects of specific neuropeptides. With an armamentarium of drugs to regulate each of the many neuropeptides, more precise modulation of feeling and thinking will be possible. The potential for alleviating emotional and neurologic disease will be great.

The ways in which current drugs interact with transmitters are best described by considering norepinephrine, one of the classic amine transmitters. Norepinephrine is a transmitter of the sympathetic nervous system, which is the part of the autonomic nervous system that prepares the body for a rapid expenditure of energy by, for example, dilating certain blood vessels, accelerating the heart rate and slowing digestion. Norepinephrine is also a brain neurotransmitter.

Neurons that contain norepinephrine, like those containing enkephalins, are highly localized in discrete areas of the brain. One of the most prominent norepinephrine pathways is centered in a small nucleus in the brainstem called the locus coeruleus, from which norepinephrine neurons send axons to many regions of the brain. In this way the relatively few neurons in the locus coeruleus are able to influence literally billions of other neurons.

There are four major types of norepinephrine receptors, called alpha₁, alpha₂, beta₁ and beta₂. The different norepinephrine receptors are at differing sites within the body, and so it is possible to design drugs that achieve their effects by blocking or stimulating the effect of norepinephrine at one particular type of receptor. For example, in the peripheral nervous system the stimulation of alpha₁ receptors raises blood pressure, and so many drugs for treating hypertension are designed to selectively block alpha₁ receptors. Stimulating beta₁ or beta₂ receptors, on the other hand, speeds the heart rate and dilates the bronchial tree of the lungs. A beta stimulant can therefore be used to treat asthma and a beta blocker to treat angina.

Unfortunately a beta stimulant administered to treat asthma might speed the heart excessively; a beta blocker intended to treat angina could worsen a case of asthma. It is possible to overcome these side effects, however, because the heart has mainly beta$_1$ receptors, whereas the lung has mostly beta$_2$ receptors. Drug designers have therefore been able to use selective beta$_2$ stimulants to relieve the symptoms of asthma without causing cardiac palpitations and beta$_1$ blockers to ease anginal distress without provoking asthma attacks. Many other drugs have been designed that take advantage of the multiplicity of receptor types that exists for most neurotransmitters.

A number of other drugs operate on a slightly different principle. After a neurotransmitter has been released and has acted on receptors, its action must somehow be terminated so that receptors can respond to the next nerve impulse. Most neurotransmitters are inactivated by a pumplike mechanism that transports the neurotransmitter back into the nerve ending from which it was released. Some drugs exert their therapeutic effects by blocking this "reuptake" mechanism. For instance, the major antidepressant drugs, which are called tricyclic antidepressants, block the reuptake inactivation of norepinephrine and serotonin. By inhibiting the norepinephrine and serotonin reuptake systems, the antidepressant drugs make a greater quantity of neurotransmitter available to receptor sites. That such drugs are indeed able to help cure depression indicates that depression may result in part from a deficiency of norepinephrine, serotonin and other biogenic amines (amine transmitters that are produced within the living brain).

Certain other drugs act by influencing the enzymes that synthesize or destroy the norepinephrine molecule. For example, by inhibiting monoamine oxidase, the enzyme that degrades norepinephrine, it is possible to cause a buildup in the levels of norepinephrine in the brain. Much of the accumulated norepinephrine is then forced out of nerve endings by osmotic pressure and acts on receptors. Monoamine oxidase inhibitors can thus increase the synaptic activity of norepinephrine, and so they can be used as antidepressants.

Communication between cells or groups of cells is crucial for the survival of every multicellular organism. In higher organisms the mechanisms of communication rely on a large number of highly specialized messenger molecules. As investigators come to know the properties and functions of all these intercellular messengers, it will become possible to develop safer and more effective therapeutic agents for conditions as diverse as hormonal abnormalities, heart disease and mental illness.

# 9

# THE MOLECULAR BASIS OF COMMUNICATION WITHIN THE CELL

# The Molecular Basis of Communication within the Cell

*The number of substances serving as signals in cells is remarkably small. Each such "second messenger" is a crucial guide for the cell, helping to determine how the cell responds to the organism's needs*

by Michael J. Berridge

The division of labor among the cells of a multicellular organism requires that each cell population be able to call on the services of some cell populations and respond to the requirements of others. Much of this coordination is achieved by chemical signals. Yet most of the arriving signals never invade the privacy of a cell. Dispersed on the outer surface of the cell are the molecular antennas known as receptors, which detect an incoming messenger and activate a signal pathway that ultimately regulates a cellular process such as secretion, contraction, metabolism or growth. The major barrier to the flow of information is the cell's plasma membrane, where transduction mechanisms translate external signals into internal signals, which are carried by "second messengers."

In molecular terms the process depends on a series of proteins within the cell membrane, each of which transmits information by inducing a conformational change—an alteration in shape and therefore in function—in the protein next in line. At some point the information is assigned to small molecules or even to ions within the cell's cytoplasm. They are the second messengers, whose diffusion enables a signal to propagate rapidly throughout the cell.

The number of second messengers appears to be surprisingly small. To put it another way, the internal signal pathways in cells are remarkably universal. Yet the known messengers are capable of regulating a vast variety of physiological and biochemical processes. The discovery of the identity of particular second-messenger substances is proving, therefore, to be of fundamental importance for understanding how cellular activity is governed. Two major signal pathways

are now known. One employs the second-messenger cyclic adenosine monophosphate (cyclic AMP). The other employs a combination of second messengers that includes calcium ions and two substances, inositol triphosphate ($IP_3$) and diacylglycerol (DG), whose origin is remarkable: they are cannibalized from the plasma membrane itself.

The paths have much in common. In both of them the initial component, the receptor molecule at the surface of the cell, transmits information through the plasma membrane and into the cell by means of a family of *G* proteins: membrane proteins that cannot be active unless they bind guanosine triphosphate (GTP). In both paths the *G* proteins activate an "amplifier" enzyme on the inner face of the membrane. The enzyme converts precursor molecules into second messengers. As a rule the precursors are highly phosphorylated, that is, rich in phosphate groups ($PO_4$). For example, the amplifier adenylate cyclase converts adenosine triphosphate (ATP) into cyclic AMP, whereas the amplifier phospholipase *C* cleaves the membrane lipid phosphatidylinositol 4,5-biphosphate, or $PIP_2$, into DG and $IP_3$.

In both paths, moreover, the final stages are similar: the second messengers induce cellular proteins to change their structure. (In one three-dimensional conformation the protein is inactive; in another it contributes to a cellular function such as secretion or contraction.) There are two main ways in which second messengers function. In one of them the second messenger acts directly. It binds to the protein (specifically, it binds to the "regulatory component" of the protein) and thus triggers a conformational change. A classic example is found in skeletal muscle. There the second messenger calcium binds to the protein troponin *C,* triggering a conformational change that leads to the contraction of the muscle. In the alternative, more common mechanism the second messenger acts indirectly: it activates an enzyme called a protein kinase, which then phosphorylates a protein. The phosphorylation (that is, the addition of a phosphate group) induces the protein to change its shape.

Of all the steps of the known second-messenger pathways the ones best understood today are the steps of transduction and amplification that ac-

**MAKING OF A MESSENGER** for the conveyance of signals inside a cell is shown in these images, which depict the structure of the precursor of the messenger and of the messenger itself. The precursor (*top*) is adenosine triphosphate, or ATP, which typically serves the cell by donating energy to chemical reactions. ATP has three parts. The nitrogenous base called adenine forms the upper right of the overall molecule; its structure is dominated by a hexagon and a pentagon of carbon atoms (*white*) and nitrogen atoms (*blue*). The adenine is joined to another pentagon, that of the sugar called ribose, seen almost end on at the bottom of the ATP molecule. In turn the ribose is linked to a chain of three phosphate groups, each consisting of a central phosphorus atom (*yellow*) and satellite oxygen atoms (*orange*). The chain wraps behind the ribose and extends toward the left. For conversion into an intracellular messenger ATP is altered into cyclic adenosine monophosphate, or cAMP (*bottom*). Two of the three phosphate groups are removed, and the remaining group becomes bound to the rest of the molecule at two of its satellite oxygen atoms. The phosphate group thus takes on a cyclic structure with respect to the rest of the AMP. The images were computer-generated from crystallographic data by Tripos Associates in St. Louis.

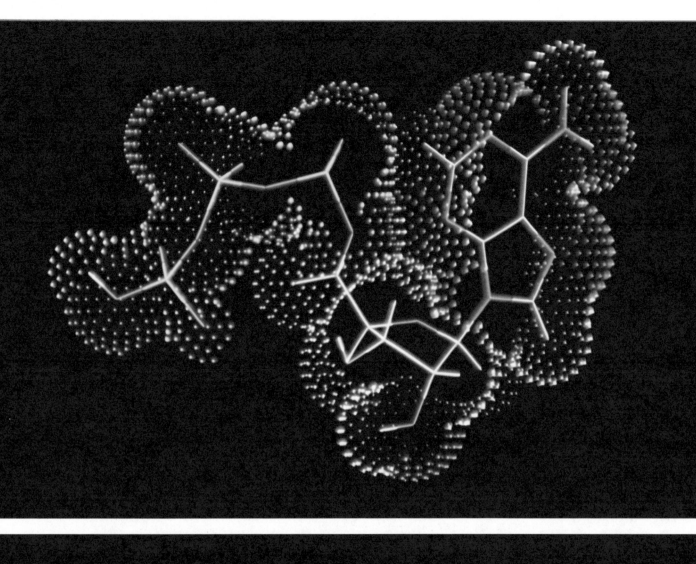
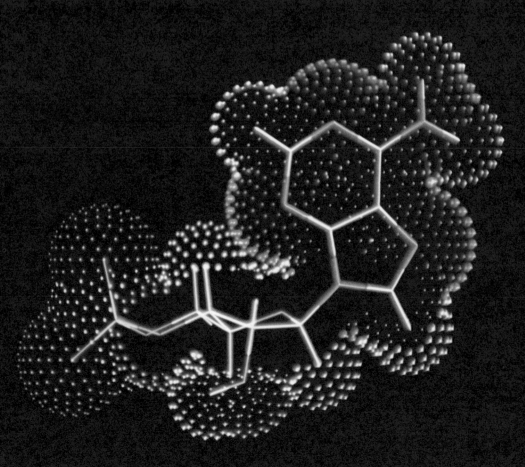

tivate cyclic AMP. The facts emerged in stages, beginning in 1958, when Earl W. Sutherland, Jr., and Theodore W. Rall, working at Case Western Reserve University, discovered cyclic AMP itself. Then in 1971 Martin Rodbell and his colleagues at the National Institutes of Health showed that GTP is essential for the transduction mechanism to generate cyclic AMP. Before information can flow across the membrane two events must occur. At the surface of the cell an external signal must bind to its receptor. Meanwhile, from inside the cell, a GTP molecule must act on its $G$ protein.

The sequence has been elucidated in detail by Alfred G. Gilman and his

colleagues at the University of Texas Health Science Center at Dallas. Two types of $G$ proteins turn out to be involved, one of them stimulatory and the other inhibitory. The stimulatory protein, called $G_s$, links itself to receptors called $R_s$. The binding of an external signal to such a receptor induces a conformational change in the receptor. The change is transmitted through the cell membrane to $G_s$, making it susceptible to GTP, which approaches from inside the cell. The binding of GTP to $G_s$ then constitutes an on-reaction: it forces $G_s$ into still another conformation, one that enables it to activate adenylate cyclase and thereby instigate the formation of cyclic AMP.

The information carried by the external signal has now been transmitted across the membrane and assigned to an internal signal: a second messenger.

The activity of the $G_s$-GTP complex is ended by the hydrolysis of the GTP to GDP (guanosine diphosphate); that constitutes the off-reaction. The hydrolysis is catalyzed by the enzyme GTPase. As it happens, the activity of GTPase is inhibited by the toxin produced by the cholera bacillus. The toxin thereby prolongs the life of the $G_s$-GTP complex, so that the cell produces cyclic AMP continually, even in the absence of an external signal calling for its manufacture. The severe diarrhea characteristic of victims of chol-

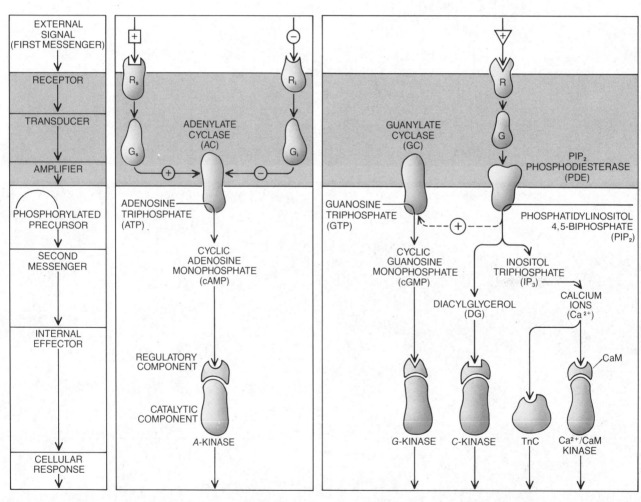

**KNOWN SIGNAL PATHWAYS IN CELLS are few in number. In functional terms they share a sequence of events (*left*). External messengers arriving at receptor molecules in the plasma membrane (*gray*) activate a closely related family of transducer molecules, which carry signals through the membrane, and amplifier enzymes, which activate internal signals carried by "second messengers." The pathway employing the second messenger cAMP (*middle*) has stimulatory receptors ($R_s$) and inhibitory ones ($R_i$), which both communicate with the amplifier adenylate cyclase (AC) by way of stimulatory and inhibitory transducers called $G$ proteins because they require guanosine triphosphate (GTP) to function. Adenylate cyclase converts ATP into cAMP. The other major pathway (*right*) is not known to recognize inhibitory external signals. It employs a stimu-** latory $G$ protein to activate its amplifier, a phosphodiesterase (PDE) enzyme. The enzyme makes phosphatidylinositol 4,5-biphosphate (PIP$_2$) into a pair of second messengers, diacylglycerol (DG) and inositol triphosphate (IP$_3$). In turn IP$_3$ induces the cell to mobilize still another messenger: calcium ions (Ca$^{2+}$). Moreover, the path somehow induces the amplifier guanylate cyclase (GC) to convert GTP into the second messenger cyclic guanosine monophosphate (cGMP). In general the second messengers bind to the regulatory component of a protein kinase, an enzyme that activates a cellular response such as contraction or secretion by adding phosphate (PO$_4$) groups to particular proteins. Calcium binds to a family of proteins including calmodulin (CaM) and troponin C (TnC). In turn CaM activates a protein kinase; TnC stimulates muscle contraction directly.

era can be explained in those terms. In the cells of the intestine cyclic AMP is a potent activator of fluid secretion.

The other type of *G* protein in the cyclic-AMP pathway mediates an inhibitory transduction. The arrival of an external signal at the receptors designated $R_i$ brings on a conformational change in the *G* protein called $G_i$ (a change again dependent on the binding of GTP); the *G* protein in turn inhibits adenylate cyclase. Here too the flow of information can be blocked by a bacterial toxin, this one produced by *Bordetella pertussis,* the causative agent of whooping cough. Pertussis toxin blocks the inhibition of adenylate cyclase; however, it is not yet known whether the blockage accounts for any of the symptoms of the disease. Bacterial toxins have been valuable experimental tools for defining the roles of *G* proteins. Another effective agent is forskolin, an organic molecule isolated from roots of the Indian herb *Coleus forskohlii.* Extracts from the plant are still employed in the folk medicine of India as remedies for ailments including heart diseases, respiratory disorders, insomnia and convulsions. Pharmacological studies have established that forskolin activates adenylate cyclase.

In the cyclic-AMP pathway the final chemical steps are mediated by an *A*-kinase: a protein kinase that phosphorylates a particular protein when it is activated specifically by cyclic AMP. Each *A*-kinase has two parts, a catalytic subunit and a regulatory subunit. Cyclic AMP binds to the regulatory subunit, thereby liberating the catalytic one, which is then free to phosphorylate proteins. In fat cells, for example, the enzyme lipase initiates the tapping of the energy content of lipid (fatty) molecules. Hormones such as epinephrine (also called adrenaline) bind to receptors on the cell surface; the receptors, acting through *G* proteins, influence adenylate cyclase, which makes cyclic AMP; the cyclic AMP stimulates an *A*-kinase, and the *A*-kinase activates lipase by phosphorylating the enzyme.

Other examples are known in which cyclic AMP works through *A*-kinase to activate cellular enzymes (or cellular processes such as ion transport). As the molecular details of the final steps in the cyclic-AMP pathway are examined in a number of cell types, however, a consistent pattern has begun to emerge: cyclic AMP often proves to serve the cell primarily by activating another second messenger, namely calcium ions. That is, one of the two known signal pathways in cells proves to act chiefly by modulating the

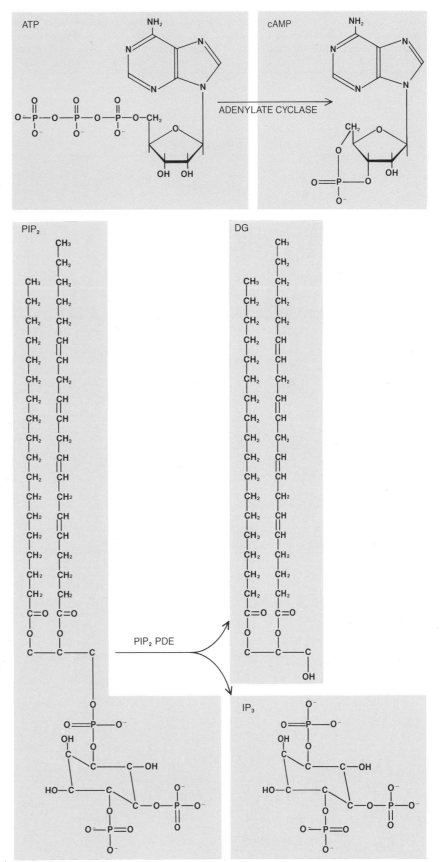

**CHEMICAL STRUCTURE** of identified second messengers is displayed for three messengers: cAMP (*also shown in illustration on page 97*), DG and IP₃. Cyclic AMP (*top*) is made in a reaction that cleaves two of the three phosphate groups from ATP. DG and IP₃ (*bottom*) are made from PIP₂ by a simple reaction: the negatively charged "head" of the precursor molecule, a phospholipid found in the inner leaf of plasma membrane, is cleaved from the glycerol backbone that carries the twin fatty acid "tails" of the precursor.

other known signal path. The heart provides a now classic example. There epinephrine acts through the cyclic-AMP pathway to modulate the intracellular level of calcium. Thus the force of each heartbeat is governed by a brief calcium pulse. Similar findings (that is, modulation of the calcium pathway by the cyclic-AMP pathway) have emerged in other muscle cells and in a variety of secretory cells, including nerve cells.

The first description of calcium as an intracellular messenger was given in 1883, when the English physician and physiologist Sydney Ringer found that the muscle tissue he was examining failed to contract when the London tap water in his tissue-culture medium was replaced with distilled water. The missing ingredient soon proved to be

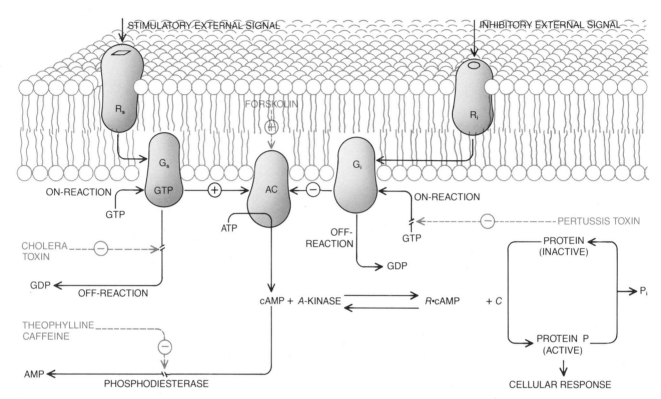

| STIMULATORY EXTERNAL SIGNALS | INHIBITORY EXTERNAL SIGNALS | TISSUE | CELLULAR RESPONSE |
|---|---|---|---|
| ADRENALINE (BETA RECEPTORS) | | SKELETAL MUSCLE | BREAKDOWN OF GLYCOGEN* |
| ADRENALINE (BETA RECEPTORS) | | FAT CELLS | INCREASED BREAKDOWN OF LIPIDS |
| ADRENALINE (BETA RECEPTORS) | | HEART | INCREASED HEART RATE AND FORCE OF CONTRACTION* |
| ADRENALINE (BETA RECEPTORS) | | INTESTINE | FLUID SECRETION* |
| ADRENALINE (BETA RECEPTORS) | | SMOOTH MUSCLE | RELAXATION* |
| THYROID STIMULATING HORMONE | | THYROID GLAND | THYROXINE SECRETION |
| VASOPRESSIN (V$_2$ RECEPTORS) | | KIDNEY | REABSORPTION OF WATER |
| GLUCAGON | | LIVER | BREAKDOWN OF GLYCOGEN* |
| SEROTONIN | | SALIVARY GLAND (BLOWFLY) | FLUID SECRETION |
| PROSTAGLANDIN I$_1$ | | BLOOD PLATELETS | INHIBITION OF AGGREGATION AND SECRETION * |
| | ADRENALINE (ALPHA$_2$ RECEPTORS) | BLOOD PLATELETS | STIMULATION OF AGGREGATION AND SECRETION * |
| | ADRENALINE (ALPHA$_2$ RECEPTORS) | FAT CELLS | DECREASED LIPID BREAKDOWN |
| | ADENOSINE | FAT CELLS | DECREASED LIPID BREAKDOWN |

**DETAILS OF SIGNAL PATHWAYS** appear in these illustrations. In the cAMP pathway (*left*) signals from stimulatory receptors (R$_s$) and inhibitory receptors (R$_i$) converge on the amplifier enzyme adenylate cyclase (AC), which converts ATP into cAMP. *G* proteins, which govern the convergence, are activated by GTP (*on-reaction*) and curtailed when the GTP is hydrolized (*off-reaction*) to GDP. For its part, cAMP binds to the regulatory component (R) of its protein kinase, liberating the catalytic component (C), which is then free to phosphorylate specific proteins that regulate a cellular response. Drugs affecting a particular stage in the sequence are shown in color. Some known cellular responses are listed in the accompanying table. In many cases (*asterisks*) cAMP proves to modulate the activity of another second messenger, calcium, which in turn governs the response. In the inositol-lipid path (*right*) external sig-

calcium. A series of observations then showed that calcium regulates not only contraction but also many other cellular processes. It is in fact the predominant second messenger in cells.

Where does the calcium come from? For certain cells, such as neurons, the source is well known: it is the extracellular fluid. Nerve signals arriving at the synaptic terminals of a neuron decrease the voltage difference across the neuronal cell membrane; the resulting "depolarization" opens voltage-sensitive calcium channels through the membrane. Before the depolarization the concentration of free calcium inside the neuron is approximately $1 \times 10^{-7}$ molar (a value corresponding to some $6 \times 10^{14}$ calcium ions per centiliter of cytoplasm). The concentration of calcium outside the neuron

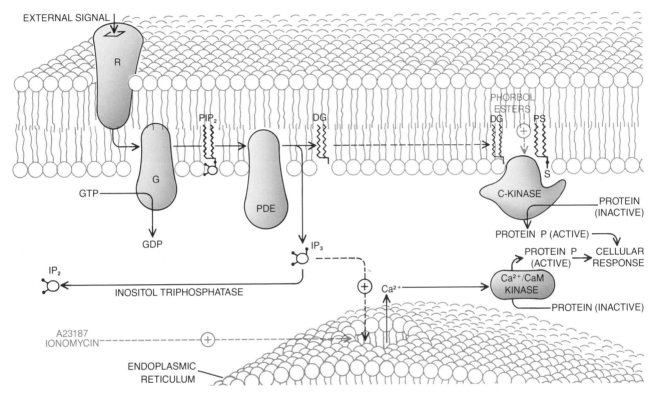

| EXTERNAL SIGNAL | TISSUE | CELLULAR RESPONSE |
|---|---|---|
| VASOPRESSIN | LIVER | BREAKDOWN OF GLYCOGEN |
| ACETYLCHOLINE | PANCREAS | AMYLASE SECRETION |
| ACETYLCHOLINE | SMOOTH MUSCLE | CONTRACTION |
| ACETYLCHOLINE | OOCYTES (*XENOPUS*) | CHLORIDE PERMEABILITY |
| ACETYLCHOLINE | PANCREATIC BETA CELLS | INSULIN SECRETION |
| SEROTONIN | SALIVARY GLAND (BLOWFLY) | FLUID SECRETION |
| THROMBIN | BLOOD PLATELETS | PLATELET AGGREGATION |
| ANTIGEN | LYMPHOCYTES | DNA SYNTHESIS |
| ANTIGEN | MAST CELLS | HISTAMINE SECRETION |
| GROWTH FACTORS | FIBROBLASTS | DNA SYNTHESIS |
| LIGHT | PHOTORECEPTORS (*LIMULUS*) | PHOTOTRANSDUCTION |
| SPERMATOZOA | SEA URCHIN EGGS | FERTILIZATION |
| THYROTROPIN RELEASING HORMONE | ANTERIOR LOBE OF PITUITORY GLAND | PROLACTIN SECRETION |

**nals bind to receptors (R), which transmit information through a $G$ protein (G) to activate $PIP_2$ phosphodiesterase (PDE). In turn PDE cleaves $PIP_2$ into the second messengers inositol triphosphate ($IP_3$) and diacylglycerol (DG). The $IP_3$ is water-soluble, and so it diffuses into the cytoplasm. There it releases calcium from storage in the membranous intracellular caverns called endoplasmic reticulum. In turn the calcium stimulates a protein kinase. The DG remains in the membrane, where it activates the enzyme $C$-kinase; the membrane phospholipid called phosphatidyl serine (PS) is a cofactor, or necessary adjunct, for the activation. The two limbs of the pathway lead to the phosphorylation of distinct sets of proteins. The limbs can be activated independently by means of the drugs indicated in color. Again an accompanying table lists some of the cellular responses known to be mediated by the pathway.**

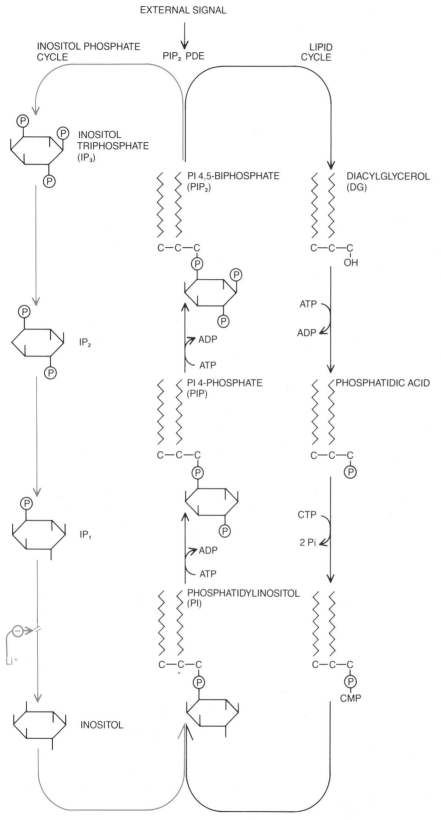

EXTERNAL SIGNAL

INOSITOL PHOSPHATE CYCLE

PIP₂ PDE

LIPID CYCLE

INOSITOL TRIPHOSPHATE (IP₃)

PI 4,5-BIPHOSPHATE (PIP₂)

DIACYLGLYCEROL (DG)

IP₂

ADP

ATP

ATP

ADP

PI 4-PHOSPHATE (PIP)

PHOSPHATIDIC ACID

IP₁

CTP

2 Pi

ADP

ATP

PHOSPHATIDYLINOSITOL (PI)

Li⁺

INOSITOL

CMP

**INOSITOL-LIPID CYCLES replenish the supply of second messengers made from inositol lipids. External signals act through the enzyme PIP₂ phosphodiesterase, which cleaves PIP₂ into the messengers DG and IP₃. The two are then directed through a sequence of chemical reactions that prepare them to be rejoined, forming phosphatidylinositol (PI) and ultimately remaking the PIP₂. The cycles require the continuous presence of ATP and cytosine triphosphate (CTP), which are sources of phosphate groups (Pᵢ). Among the few drugs known to affect a part of the path, lithium (*color*) is notable: it blocks the conversion of IP₁ into the free inositol required for the synthesis of PI. Accordingly lithium may turn out to interfere with signal pathways employing the messengers derived from inositol lipids.**

is four orders of magnitude greater. The depolarization enables calcium ions to flood into the neuron and trigger a cellular response. Even a rather small change in the intracellular concentration of calcium can exert profound changes in cellular activity. In the synaptic terminals of neurons, for example, calcium induces the release of neurotransmitter molecules.

The extracellular fluid cannot, however, be the sole source of calcium ions. For one thing, the absence of extracellular calcium does not prevent the external messenger acetylcholine from stimulating the pancreas to release the digestive enzyme amylase. Thus it has slowly become apparent that the calcium employed by a cell for internal signaling not only enters the cell from outside but also is released from internal reservoirs. There turn out to be many examples of hormones or neurotransmitters employing internal calcium to control physiological processes. The external signal gains access to the internal calcium by stimulating the hydrolysis of a lipid molecule that is part of the plasma membrane.

In 1953 Mabel N. and Lowell E. Hokin, working at the Montreal General Hospital, observed that the administration of acetylcholine to secretory cells of the pancreas increased the incorporation of radioactive phosphate groups (PO₄ groups containing phosphorus 32) into phosphatidylinositol (PI), one of the phospholipid constituents of cell membranes. Like other membrane lipids it has a hydrophobic part (two fatty acid chains attached to glycerol) linked to a hydrophilic part, in this case the "head group" inositol phosphate. Stimuli such as acetylcholine cause it to be cleaved into these two components. The increased incorporation of phosphorus 32 observed by the Hokins was a secondary event due to the subsequent resynthesis of PI.

The key point was that an external signal had been found to stimulate the turnover (the hydrolysis and resynthesis) of a membrane lipid. The Hokins proposed that the increased turnover had something to do with the mechanism of exocytosis by which the cells of the pancreas release digestive enzymes. Subsequent studies established a broader conclusion: the increased turnover occurred in response to many stimuli, not necessarily the ones that activate secretion. Hence the impression grew that the turnover of membrane lipids plays a more general role in the life of cells.

In 1975 Robert Michell of the University of Birmingham suggested such a role. Noting a strong correlation between the ability of an external signal to stimulate inositol-lipid turnover and the mobilization of calcium inside the cell, he suggested that the change in lipid turnover triggered by external signals is responsible for generating internal calcium signals. Michell and his colleagues made a further suggestion. Cell membranes contain three inositol lipids, but only one of them, the relatively minor inositol lipid phosphatidylinositol-4,5-biphosphate, or $PIP_2$, seemed to change (in particular, it was hydrolyzed) as part of the mechanism mobilizing calcium.

How might the hydrolysis of a particular membrane lipid act to increase the intracellular concentration of calcium? It has taken almost a decade to work out the details. In order to understand how second messengers can emerge from cell membrane and how they might function, a digression is required: I must turn to some basic aspects of the biochemistry of the molecules composing cell membranes. Placed in an aqueous medium, phospholipids, of which the inositol lipids are notable examples, spontaneously coalesce into the orderly double-layered alignment that constitutes the basic structure of a biological membrane. The traditional view is that the lipid bilayer functions as an inert, permeable barrier. Hence phospholipids have often been dismissed as playing a rather passive role in the life of the cell. A role for a membrane phospholipid in an intracellular signal pathway comes, therefore, as something of a surprise.

Phosphatidylinositol is a typical phospholipid, situated primarily in the inner leaflet of the bilayer. The remarkable thing about it is that it gets converted into $PIP_2$, an unusual phospholipid that has three phosphate groups instead of the one group found in all other membrane lipids. The additional phosphates, derived from ATP, are added sequentially and specifically to the carbon-4 and -5 positions of the six-carbon ring in inositol.

$PIP_2$ is the inositol lipid that interested Michell. In response to external signals it is hydrolyzed into diacylglycerol and inositol triphosphate, or $IP_3$. Two groups of investigators, Richard Haslam and Monica Davidson of McMaster University in Ontario and Shamshad Cockcroft and Bastion D. Gomperts of University College London have found that GTP is a necessary part of the sequence. Again, therefore, it seems that a $G$ protein couples surface receptors to an amplifier (in this case the enzyme phospholipase $C$). Ultimately the diacylglycerol and $IP_3$ are recycled, the first by a series

of reactions composing what is called the lipid cycle, the second by a series of reactions called the inositol phosphate cycle. The two cycles merge, reconstituting phosphatidylinositol.

The final step in the inositol phosphate cycle is particularly interesting; it is the conversion of inositol monophosphate ($IP_1$) into free inositol by the enzyme inositol 1-monophosphatase. Working in St. Louis at the Washington University School of Medicine, James Allison and William R. Sherman have shown that the action of the enzyme is inhibited by lithium ions. The administration of lithium may therefore slow the rate of resynthesis of phosphatidylinositol, impairing the effectiveness of any neuronal mechanisms that depend on the inositol lipids to carry signals. Perhaps that accounts for the efficacy of lithium ions in controlling manic-depressive mental illness.

It was while measuring the rate at which inositol phosphates form in the salivary gland of the fly in response to the hormone serotonin that my attention was first drawn to $IP_3$. Analysis of the water-soluble metabolites extracted from the gland revealed the presence of at least four distinct substances. One was free inositol; the others turned out to be the inositol phosphates $IP_1$, $IP_2$ and $IP_3$. The analysis was done in collaboration with Robin Irvine and Rex Dawson of the Agricultural Research Council's Institute of Animal Physiology in Babraham and Peter Downes of the Medical Research Council's Neurochemical Pharmacology Unit in Cambridge. In 1964 Dawson, who had been studying the enzyme that hydrolyzes $PIP_2$, stored some of the reaction product, $IP_3$, in a deep freeze at Babraham. Almost two decades later it served as a standard by which to verify our identification of the fly-gland metabolite.

By comparing the rate of $IP_3$ formation with the rate of onset of physiological signs that accompany the secretion of fluid by cells of the insect salivary gland, John P. Heslop and I were able to show that the administration of serotonin brings on the generation of $IP_3$ at a rate fast enough so that $IP_3$ could conceivably function as a second messenger: it would mobilize calcium, which in turn would cause the secretion of saliva. Seeking direct evidence of such a function, we sent a sample of $IP_3$, which Irvine had prepared from red blood cells, to the Max Planck Institute in Frankfurt. There Hanspeter Streb and Irene Schulz found that when the sample was applied to cells from the pancreas of a rat, it caused a profound release

of calcium. (The cells had first been permeabilized so that the $IP_3$ applied by the investigators could gain access to the interior of the cells.)

This first demonstration of a release of calcium induced by $IP_3$ has now been confirmed in a number of different types of cells. Gillian Burgess and James W. Putney, Jr., of the Medical College of Virginia and John R. Williamson of the University of Pennsylvania School of Medicine have shown, for example, that $IP_3$ mobilizes stored calcium as part of the hormonal mechanism for releasing glucose from the liver. Moreover, Yoram Oron and his colleagues at Tel-Aviv University have shown that the current of chloride ions induced by acetylcholine in immature oocytes, or egg cells, of the frog Xenopus can be duplicated by injecting the cells with $IP_3$. In addition $IP_3$ elicits many of the early events of fertilization. For example, granules stored just under the surface of the egg are normally secreted within minutes of fertilization to form the thick protective layer called the fertilization membrane. The secretion, which is calcium-dependent, can be triggered by injecting eggs with $IP_3$. In each case the $IP_3$ proves to act predominantly by causing the release of calcium from imprisonment in the cell's endoplasmic reticulum, an internal membrane that forms a system of caverns inside the cell. In turn the calcium elicits the cellular response.

The discovery of the second messenger $IP_3$ has led to speculation that $IP_3$ may function as a second messenger in skeletal muscle. In muscle the depolarization of the infoldings of muscle membrane known as transverse tubules somehow triggers the release of calcium from the sarcoplasmic reticulum, a structure analogous to the endoplasmic reticulum of nonmuscle cells. The calcium triggers muscle contraction. Roger Y. Tsien and Julio Vergara of the University of California at Berkeley and Tullio Pozzan and his colleagues at the University of Padua have found that isolated muscle fibers contract in response to the administration of $IP_3$. The idea, then, is that in muscle $IP_3$ functions as the link between depolarization and calcium. The verification of the speculation would be the jewel in the crown of the $IP_3$ hypothesis; the problem of how calcium signals are generated in skeletal muscle has puzzled physiologists for decades.

A further reason the inositol-lipid transduction mechanism is attracting much interest is that the signal pathway bifurcates. One product of the hydrolysis of the inositol lipid $PIP_2$

is IP$_3$, whose role I traced above. The other product, diacylglycerol, remains in the membrane, yet apparently it functions, like IP$_3$, as a second messenger. Yasutomi Nishizuka and his colleagues at Kyoto University propose that it activates a membrane-bound protein kinase, which they have named *C*-kinase.

The contribution of each limb of the bifurcating inositol-lipid path can be assessed with pharmacological agents that mimic the action of a particular second messenger and therefore stimulate only one limb of the path. The phorbol esters (substances found in the oil expressed from the seed of the small Southeast Asian tree *Croton tiglium*) mimic the action of diacylglycerol by acting directly on *C*-kinase. (The phorbol esters cause inflammation of the skin and are potent tumor-inducing agents when they are applied to experimental animals in combination with a carcinogen.) On the other hand, calcium ionophores (molecules that shield the electric charge of a calcium ion and smuggle it across the cell membrane) mimic the action of IP$_3$ by introducing free calcium into the cell. The studies establish that the two limbs are synergistic: in blood platelets, for example, Nishizuka has found that the combina-tion of a phorbol ester and a calcium ionophore induces a maximal secretion of serotonin at doses that have no effect when each drug is administered alone.

The importance of the overall two-branched signal pathway is hard to overstate: a great many cellular processes can be switched on experimentally by the combined administration of a phorbol ester and a calcium ionophore. Perhaps the most notable finding is that the synthesis of DNA can be initiated: the finding is notable because it suggests that the signal pathways responsible for routine cellular activities such as secretion and contraction may also regulate growth. The action of phorbol esters as tumor-promoting agents is probably based, for example, on their ability to amplify the DG/*C*-kinase limb of the inositol-lipid signal pathway. Indeed, the prospect arises that alterations of intracellular signal pathways may be a cause of cancer, in which the normal regulation of cell growth is disrupted.

Cells grow by progressing through the stages of the cell cycle. In an initial phase they increase in size. This is the first growth phase, G$_1$. Next they replicate all their chromosomes (during S, the DNA-synthesis phase) and prepare themselves for cell division (during G$_2$, the second growth phase). Finally they divide (during M, the mitosis phase). Just after cell division comes a branch point. Each daughter cell arising from the division can reenter the cell cycle and hence can divide again. Alternatively, a daughter cell can enter the G$_0$ phase, during which it differentiates, becoming capable of performing some specialized task in one of the tissues of the body. For certain types of cells, such as neurons, differentiation puts an end to division; for other types the cell has the option of returning to the cell cycle to engender further progeny. In the latter case the return to the cell cycle is determined by the action of growth factors: substances released by one group of cells that stimulate the growth of others.

Just how growth factors act is still very much a mystery. Clearly, however, they must instigate the sending of signals from the cell surface (where the factors act by binding to receptors) to the nucleus (where DNA is replicated). I shall consider two possible pathways. One of them is not yet well understood. Employed by growth factors such as insulin and epidermal growth factor (EGF), it appears to

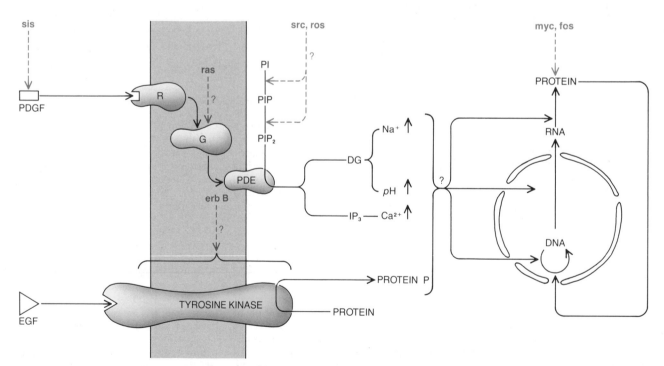

**REGULATION OF CELL GROWTH is presumably a function of second-messenger pathways. The central problem, therefore, is to explain how the external signals called growth factors, which act on cell-surface receptors (*left*), can instruct the machinery in the nucleus (*right*) to begin the complex sequence of events that culminates in DNA synthesis. Some growth factors, including platelet-derived growth factor (PDGF), appear to employ the conventional inositol-lipid pathway: IP$_3$ mobilizes calcium, whereas DG activates a membrane-bound pump that exchanges protons for sodium ions, thus increasing intracellular *p*H and the concentration of so-dium ions. Each change could convey information to the nucleus. Other growth factors, including epidermal growth factor (EGF), appear to employ a different pathway. The EGF receptor, which spans the plasma membrane, includes an inner part that phosphorylates proteins (it shows tyrosine kinase activity). A number of oncogenes (genes whose inappropriate function is linked to cancer) may act (*color*) by disrupting the growth-control pathways. The *erb B* oncogene specifies a protein identical with part of the EGF receptor. Several other oncogenes appear to affect the inositol-lipid pathway. Question marks indicate the more speculative possibilities.**

rely on receptors that activate the enzyme tyrosine kinase. The pathway may be in essence a cascade based on the phosphorylation of a succession of proteins.

The other signal pathway, employed by such factors as platelet-derived growth factor (PDGF), appears to be identical with the pathway employed by hormones and neurotransmitters. The PDGF arriving at the surface of a cell stimulates the hydrolysis of $PIP_2$ into the second messengers $IP_3$ and diacylglycerol, which may then contribute to the events that make up growth phase $G_1$ and prepare the cell for DNA synthesis. More specifically, $IP_3$ seems to act by mobilizing intracellular calcium; diacylglycerol activates $C$-kinase, which in turn activates a membrane-bound ion-exchange mechanism. The mechanism extrudes protons (hydrogen ions) from the cell, thus raising intracellular $pH$. Together the activation of calcium and the raising of $pH$ are thought to contribute to the synthesis of RNA and protein that prepares the cell to synthesize DNA.

Since each signal pathway consists of a sequence of reactions controlled by specific proteins (receptors and enzymes), the genetic material of the cell must include the genes responsible for the synthesis of the proteins the pathways require. Any aberration of the function of such genes might lead, therefore, to abnormalities of cellular growth, and conceivably to the uncontrolled growth and structural transformations typical of cancer. About 25 remarkable genes have in fact been identified: genes whose inappropriate function has been linked to the incidence of cancer. They are collectively termed oncogenes. Until recently the precise normal function of each such gene was obscure. It is now apparent that some of them encode the structure of various components of the signal pathways controlling cell growth.

The first link between an oncogene and a component of an intracellular signal pathway was made simultaneously by two groups, one headed by Russell F. Doolittle of the University of California at San Diego, the other by Michael Waterfield of the Imperial Cancer Research Fund Laboratories in London. The groups discovered that the oncogene called *sis* controls the synthesis of platelet-derived growth factor. Discoveries of similar import followed at other laboratories. The *erb b* gene proved to encode the structure of a protein almost identical with the epidermal-growth-factor receptor. The receptor has three main parts. An external part, exposed at the surface of the cell, includes the EGF-binding domain; a middle part spans the cell membrane; an inner part, exposed to the cytoplasm, expresses the protein-phosphorylating activity of a tyrosine kinase. The product of the *erb b* gene is a truncated version of the receptor, a version that lacks the external part of the protein. Conceivably the truncated version initiates signals inside the cell even in the absence of EGF.

The *ras* oncogene also fits the pattern. It is known to be active in many types of cancer cell. Its function is not yet known, but its product has characteristics of a $G$ protein: it is a constituent of the cell membrane and it binds and hydrolyzes GTP. One possibility is that it may intervene between growth-factor receptors and such signal amplifiers as the phosphodiesterase enzyme that cleaves $PIP_2$ into the second messengers $IP_3$ and diacylglycerol. Two oncogenes, *src* and *ros*, seem to have a role in the conversion of phosphatidylinositol into $PIP_2$. That is, they appear to regulate the enzymes that replenish the precursor of the inositol-lipid second messengers. Two other oncogenes, *myc* and *fos*, apparently function at the other end of the intracellular signal cascade. Philip Leder and his colleagues at the Harvard Medical School have found that the abundance of messenger RNA transcribed from the *myc* oncogene increased greatly within an hour of treating fibroblasts (immature connective-tissue cells) with PDGF. Transcripts of the *fos* oncogene appear even earlier. The *myc* and *fos* genes specify the structure of proteins found in the nucleus and so may prove to take part in the sequence of events initiating the synthesis of DNA.

One begins to see that an integrated network of oncogene products is responsible for conveying information from the cell surface to the nucleus. Some oncogenes (*sis*) specify growth factors, which act by inducing other oncogenes (*myc* and *fos*) to produce substances active within the nucleus. Distortions of such sequences lead to uncontrolled cell growth and cancer.

In all the foregoing I have neglected a further second-messenger candidate. It is cyclic GMP, which differs structurally from cyclic AMP in that guanosine takes the place of adenosine. Although cyclic GMP has the hallmarks of a second messenger, its precise role in the cell is not well understood. In the first place, guanylate cyclase, the enzyme that makes cyclic GMP from GTP, is usually not connected to a receptor. Nevertheless, the formation of cyclic GMP often occurs together with the activation of the inositol-lipid pathway. Apparently some molecule created by the hydrolysis of inositol lipids brings on the formation of cyclic GMP. The end of the signal pathway is equally obscure. Cyclic GMP is known to activate a protein kinase (in particular the one called $G$-kinase), which in turn phosphorylates certain proteins. Their functions are not known.

Still, cyclic GMP has some striking effects, which have been demonstrated best in the nervous system. For example, James W. Truman of the University of Washington has uncovered a role for cyclic GMP in controlling a complex pattern of insect behavior. At the end of metamorphosis moths escape from their cocoons by means of a carefully orchestrated pattern of writhing and wriggling triggered by an eclosion (hatching) hormone released from the brain. This preprogrammed behavioral pattern is initiated by an increase in the level of cyclic GMP, which occurs when eclosion hormone acts on the nervous system.

Another tissue in which a clear function for cyclic GMP is beginning to emerge is the retina. Specifically, the function is emerging in the vertebrate photoreceptors, or light-sensitive cells, known as rod cells. A rod cell is a sensory transducer stationed between the visual world and the brain. It is an elongated cell. At one end it receives photons, or quanta of light; at the other end it releases a neurotransmitter, thus dispatching signals to neurons. Two things about the sequence are remarkable. First, the amount of transmitter released by the cell is greatest in the absence of light. This suggests a curious attribute of rod cells: in response to external signals (in this case photons) the internal messenger must decrease its activity. Second, the receptor in rod cells is the molecule rhodopsin, which is present in an elaborate stack of disks of membrane inside the cell. On the other hand, the release of neurotransmitter is regulated by changes in the permeability of the plasma membrane to sodium ions. That defines the central problem in phototransduction: What is the identity of the second messenger that carries information from the internal disks to the plasma membrane?

For some time two camps of investigators have held opposing views about the identity of the messenger. One camp championed calcium; the other championed cyclic GMP. The truth may lie in the middle: both may take part. Here I shall concentrate on cyclic GMP. The current hypothesis is that sodium channels through the plasma membrane of the rod cell are kept open in the dark by a high intracellular level of cyclic GMP. A surprising aspect of the hypothesis has been reported by Evgenii Fesenko and his colleagues at the Institute of Biological Physics in Moscow. Cyclic GMP appears to open the channel directly, without activating a protein kinase. When photons arrive, they are absorbed by rhodopsin, which in response induces a molecule called transducin (another member of the *G*-protein family) to bind GTP and activate the enzyme called cyclic-GMP phosphodiesterase. The result is a precipitous fall in the level of cyclic GMP and the closure of sodium channels. It should be noted, however, that in the photoreceptors of the crab *Limulus* the effect of light can be duplicated by the injection of $IP_3$ but not of cyclic GMP. Perhaps the second messenger in visual signal transduction varies from species to species.

Interest in cyclic GMP is likely to grow now that Ferid Murad of the Stanford University School of Medicine has shown that atrial natriuretic factor, a newly discovered hormone secreted by the atrium of the heart, seems to relax the smooth muscle surrounding blood vessels (and so take part in regulating blood pressure) by increasing the level of cyclic GMP. Like cyclic AMP, cyclic GMP may act by modulating the action of calcium.

Have all the second-messenger signal pathways been identified? The answer is almost certainly no. There are external signals that induce profound effects in cells by way of a signal pathway that remains totally obscure. An intriguing example is offered by the maturation of starfish oocytes. The administration of 1-methyl adenine (the maturation hormone released from the surrounding follicle cells) to the surface of starfish oocytes causes dissolution of the nucleus and the reinitiation of meiosis: the process of cell division by which sex cells such as oocytes increase in number. Just how a simple substance such as 1-methyl adenine acting at the cell surface can make the nucleus disappear is a mystery. Another example is offered by insulin. The pathway by means of which it drives lipid and glycogen synthesis in muscle and liver cells is a mystery. Joseph Larner and his colleagues at the University of Virginia School of Medicine have proposed that insulin may act through a peptide (a short amino acid chain), but the evidence is far from complete. All that seems certain is that the insulin receptor, like the receptor for epidermal growth factor, acts as a tyrosine kinase. Although some of the signal pathways in cells have now been mapped out, it seems clear that uncharted pathways remain.

**10**

# THE MOLECULAR BASIS OF DEVELOPMENT

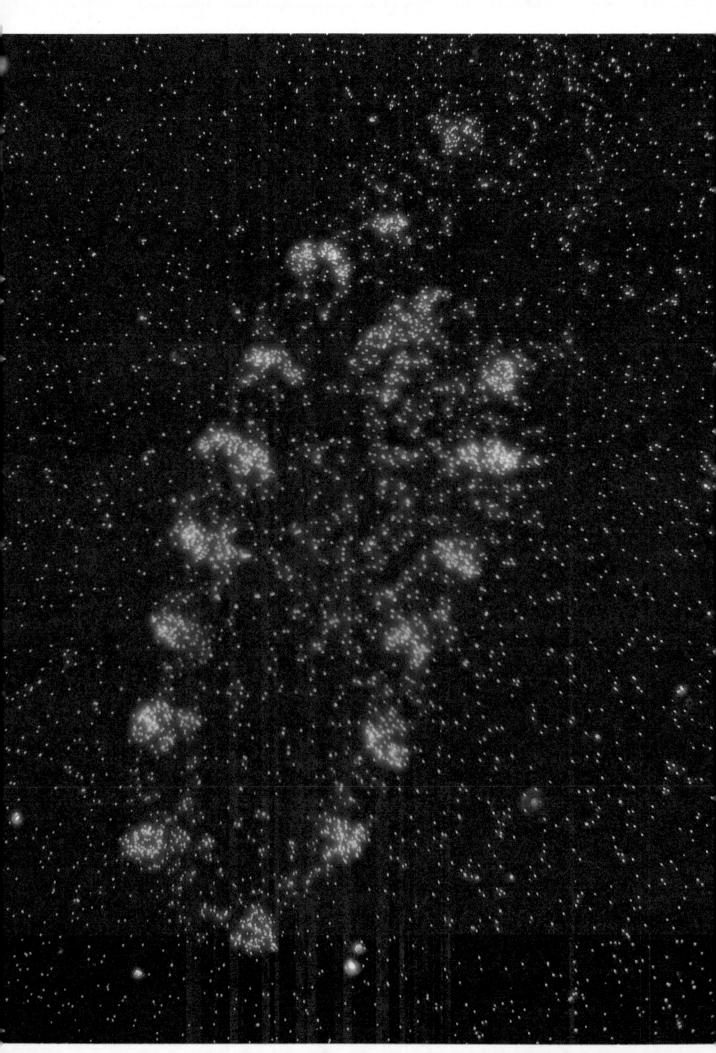

# The Molecular Basis of Development

*How is the basic architecture of an embryo laid down? The discovery
of a short stretch of DNA called the homeobox apparently provides a
crucial part of the answer in a remarkably wide range of organisms*

by Walter J. Gehring

How the linear information contained in the DNA can generate a specific three-dimensional organism in the course of development from the fertilized egg is one of the great mysteries of biology. Each organ of the mature animal carries out a particular task and consists of specialized tissues; the tissues in turn consist of specialized cells. Such specialized cells display only a small fraction of the great genetic potential of the fertilized egg. What fraction of the full potential a particular cell displays depends on which of its genes are turned on and which are turned off. Each cell

is characterized by a specific pattern of active and inactive genes that undergoes sequential changes as development proceeds. Since the genome (the full genetic complement) of a higher organism may include as many as 50,000 genes, it is unlikely that each gene is regulated individually. It must be assumed that the genes are regulated in groups, with a "master" gene controlling the action of each group. Although such a scheme has long seemed plausible, it proved difficult to find the master genes.

In the past five years some of the master genes that control development

have been identified. Clearly, if development is to proceed correctly, both the timing of developmental events and the spatial organization of tissues in the embryo must be regulated precisely. The recently identified genes affect both functions. A group of genes found in the transparent roundworm *Caenorhabditis elegans* apparently play a crucial role in the timing of differentiation of the cells in that organism. The most intriguing findings, however, concern the spatial organization of the embryo, and they come from work on the fruit fly *Drosophila melanogaster*. Using the new methods of molecular biology, my colleagues and I found that many of the genes controlling spatial organization in *Drosophila* have a common segment of DNA. That segment, which has been named the homeobox, may enable the genes that contain it to regulate the activity of batteries of other genes.

When the gene containing the homeobox is translated into a protein, the homeobox yields a stretch of amino acids that is thought to bind to the DNA double helix. By binding to the DNA of particular genes the protein may be able to turn them on or off. If the appropriate battery of genes were turned on in one group of cells in the *Drosophila* embryo, those cells might be directed down the pathway that leads to becoming part of a wing; the activation of a different battery of genes in a second group of cells might send those cells toward becoming part of a leg. The significance of the homeobox, however, extends beyond *Drosophila*. The common DNA sequence has now been found in a range of organisms that extends from worms to human beings. Perhaps the homeobox will prove to be the key to uncovering the mechanisms of development in the higher organisms. Whether those high hopes are fulfilled or not, it is apparent that in the past five years developmental biology has entered a phase in

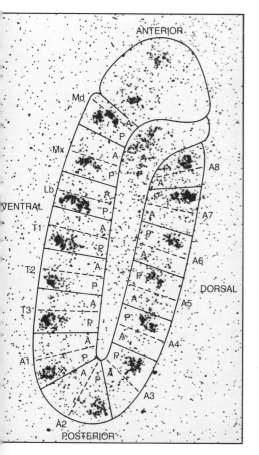

**PATTERN OF EXPRESSION for a gene
called** *engrailed* **is shown by the distribution
of 14 bright spots in a thin section of an embryo of the fruit fly** *Drosophila melanogaster*. **The bright spots indicate the presence
of** *engrailed* **transcripts (messenger-RNA
molecules incorporating the information in
the gene), which accumulate at the sites in
the embryo where the gene is expressed. The
section shown is a longitudinal one through
an embryo six hours after fertilization. At
that stage the embryo has already been divided into segments characteristic of the
fly's body plan (key at left). There are at
least three head segments (Md, Mx, Lb),
three thoracic segments (T1–T3) and eight
complete abdominal segments (A1–A8).
(The abdominal segments, which will ultimately form the posterior end of the insect,
have migrated over the dorsal surface in a
complex developmental movement.) Each
segment is divided into an anterior compartment (A) and a posterior one (P).** *Engrailed*
**is expressed in the 14 posterior compartments and helps to endow them with their
identities. The** *engrailed* **transcripts were
detected by the technique called in situ
hybridization. Radioactively labeled DNA
from the** *engrailed* **gene was applied to a
section of the embryo. The DNA hybridized, or bound selectively, to** *engrailed* **messenger RNA (mRNA). The tissue was covered with photographic emulsion and developed. Each bit of radioactive DNA in
a DNA-RNA hybrid exposed silver grains
in the emulsion, yielding a bright speck.**

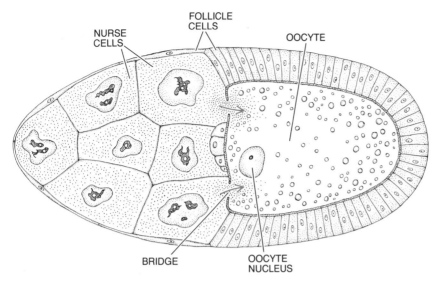

NURSE CELLS

FOLLICLE CELLS

OOCYTE

BRIDGE

OOCYTE NUCLEUS

**DROSOPHILA OOCYTE, or unfertilized egg, is formed within an ovarian structure called the follicle; the illustration shows the follicle in cross section. The oocyte and 15 nurse cells are formed from the same precursor cell by four rounds of cell division. Cytoplasmic bridges enable the nurse cells to transfer RNA, proteins and cellular organelles to the egg. Among the substances transferred in this way may be molecules that establish the initial spatial polarity of the zygote (fertilized egg) in the early stages of embryogenesis.**

which explanations can be sought on the molecular level.

That genes controlling the timing of events in embryogenesis were discovered in *C. elegans* is due partly to the small nematode's developmental plan. The embryo of *C. elegans* develops in a way that renders the chronological pattern of events particularly significant. The cells of the *C. elegans* embryo display little flexibility. The worms have a fixed number of cells and the fate of almost every cell is determined by its ancestry. Each of the 959 cells in the adult can be traced back through the lineage of its predecessors to the fertilized egg. The pattern of these cell lineages is essentially invariant from one worm to the next, unless the animal has been affected by a mutation. In order to give rise to specialized structures in the adult the lineages must branch off at precisely the right time.

*C. elegans* was introduced to the laboratory in the 1960's by Sydney Brenner of the Medical Research Council's Laboratory of Molecular Biology in Cambridge. Since then several investigators have discovered genes that affect when the cell lineages are assigned their fates. Such genes, which are called chronogenes, have been discovered by observing the effect of mutations in them. Specific mutations in chronogenes alter the development of cell lineages, causing them to branch off sooner or later than in the normal organism. *C. elegans* normally passes through four larval stages and a single

adult stage. At the end of each stage the worm molts and thereby acquires a new cuticle, or outer covering. A particular mutation in the gene called *lin-14* causes the worm to undergo two extra molts. During the supernumerary stages the development of the cuticle is retarded: although the animal is sexually mature, the cuticle is larval.

From the effect of such mutations it can be inferred that the function of the wild-type, or normal, *lin-14* chronogene is to ensure that the cuticle-forming cells differentiate to produce the adult cuticle at the right time. Other chronogenes have been found that influence the differentiation of many other cell lineages. Control of the timing of developmental events is of particular importance in an organism such as *C. elegans*, whose developmental plan relies on invariant cell lineages. Most organisms, however, do not develop in this way. Indeed, *C. elegans* falls at one extreme of the spectrum of developmental plans. At the other end of the spectrum are organisms such as the mouse whose embryonic cells retain considerable flexibility. An early mouse-embryo cell may end up in almost any adult structure. Its fate is determined not by its place in a fixed lineage but by the spatial position it happens to take up in the early embryo.

Most organisms fall between these extremes. Among them is *Drosophila,* which has provided the lion's share of all current knowledge about the genetic control of development. Thomas Hunt Morgan of Columbia

University brought *Drosophila* into the laboratory at the beginning of this century and utilized the fly to demonstrate the chromosomal basis of heredity and the linear arrangement of genes on chromosomes. Certain features of *Drosophila* render it uniquely valuable for the study of heredity. Among them are the gigantic polytene chromosomes present in many cells, particularly those of the salivary glands. Whereas the chromosomes of most organisms include one copy of each gene, the polytene chromosomes of *Drosophila* can include as many as 1,000 copies of each gene, lined up side by side like matchsticks. As a result of this stupendous multiplicity, individual genes can be stained and seen in the light microscope as dark bands. Furthermore, *Drosophila* has many progeny, a short generation time and a relatively small genome.

The development of an individual *Drosophila* starts in the ovary of the female, where a primitive germ cell begins a highly specialized pattern of cell division. The germ cell divides four times, yielding 15 nurse cells and the oocyte, which later becomes the egg. Within a structure called the follicle the nurse cells nourish the egg. Proteins, RNA molecules and organelles such as mitochondria pour through channels that connect the nurse-cell with the egg. The contribution of the nurse cells helps to construct the egg and prepare it for fertilization. At fertilization the egg and sperm unite and the resulting zygote contains chromosomes from both parents. Almost immediately the zygote nucleus divides and its daughter nuclei begin a series of rapid divisions, one every 10 minutes. Every time the nuclei divide the quantity of DNA in the cell must be doubled. That formidable task keeps the nuclei of the zygote fully engaged and few zygotic genes are expressed.

In the usual mode of cell division the nuclei are separated by newly formed cell membranes. The daughter nuclei of the *Drosophila* zygote, however, continue to share a common cytoplasm as they divide. During the early divisions the nuclei are scattered through the cytoplasm. After the eighth round of division, when there are 256 nuclei, the nuclei begin to migrate toward the surface of the egg into the layer called the cortex. Having entered the cortical cytoplasm, the nuclei distribute themselves at the periphery in a layer one nucleus thick. Only at this stage are a substantial number of zygotic genes activated. After 13 cycles of division, cell membranes begin to divide the common cytoplasm and cells are formed. The resulting cellular monolayer is called the blastoderm.

During the next few hours the embryo develops rapidly, undergoing a complex spatial reorganization. In the process called gastrulation the interior cell layers are formed. Perhaps the most striking feature of the spatial transformation, however, is the division of the embryo into segments corresponding to the segments of the adult insect. In addition to at least three segments that are later retracted into the head (referred to as Md, Mx and Lb) there are three thoracic segments (T1–T3) and eight complete abdominal segments (A1–A8).

One day after fertilization the embryo hatches and becomes a larva that retains the segmental design of the embryo. The larva molts twice, pupates and metamorphoses into the adult fly. The adult fly is also organized in segments, but the tissues of the adult epidermis are not derived directly from the outer covering of the larva. Instead the adult epidermis arises from the imaginal disks, which are small pouches of epithelium in the body of the larva. At metamorphosis the pouches fold outward and differentiate into adult structures. For example, on each side of the body one disk gives rise to the eye and the antenna, three disks give rise to the legs, which are attached to the three thoracic segments, and one disk gives rise to the wing and a large part of the middle thoracic segment, to which the wing is attached. Thus the imaginal disks serve as building blocks for the assembly of the adult body.

The ease with which the *Drosophila* genome can be manipulated has made it possible to trace cells from adult structures back to their origin in the embryo and find out when groups of cells become committed to particular fates. Carefully controlled doses of X rays can be used to induce mutations and genetic recombinations (exchanges of DNA between chromosomes) in *Drosophila* embryos. The effects of such genetic aberrations serve as markers of particular cells or nuclei. The marked units can be followed to the adult fly to find out when they become determined: committed to give rise to a particular part of the larva or adult. Using such methods, it has been shown that prior to the formation of the blastoderm the nuclei retain complete flexibility. A nucleus can colonize any portion of the cortical cytoplasm, and the descendants of a preblastoderm nucleus can be found in

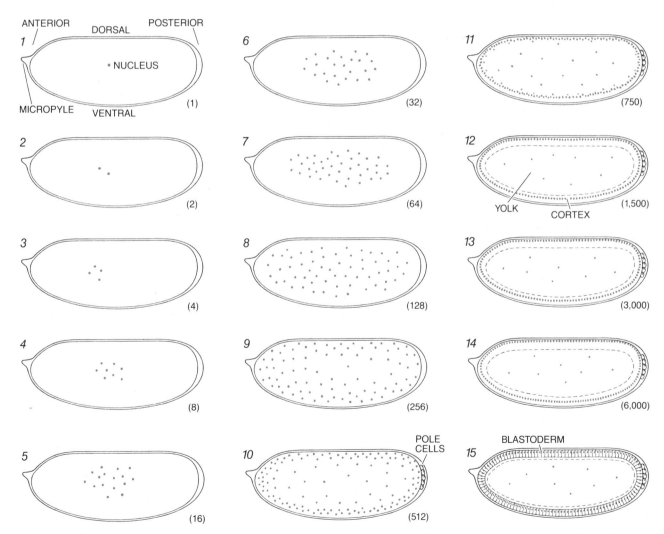

**EARLY STAGES** in the development of a *Drosophila* embryo entail repeated divisions of nuclei in a common cytoplasm. The sperm enters the egg through a structure called the micropyle, which is at the anterior end of the oocyte (*1*). The sperm nucleus unites with the egg nucleus. Synchronized cycles of nuclear division beginning soon after fertilization double the number of nuclei about every 10 minutes (*2–8*). After the eighth division, when there are 256 nuclei, the nuclei begin to migrate to the cortex, or periphery, of the egg (*9*). When there are 512 nuclei, the first cell membranes are formed around a group of cells at the posterior end of the egg (*10*). These "pole cells" ultimately give rise to the germ cells of the adult fly. The other nuclei continue to divide in the cortex (*11–14*). When there are about 6,000 nuclei in the periphery, membranes separate the nuclei into a monolayer of cells called the blastoderm (*15*).

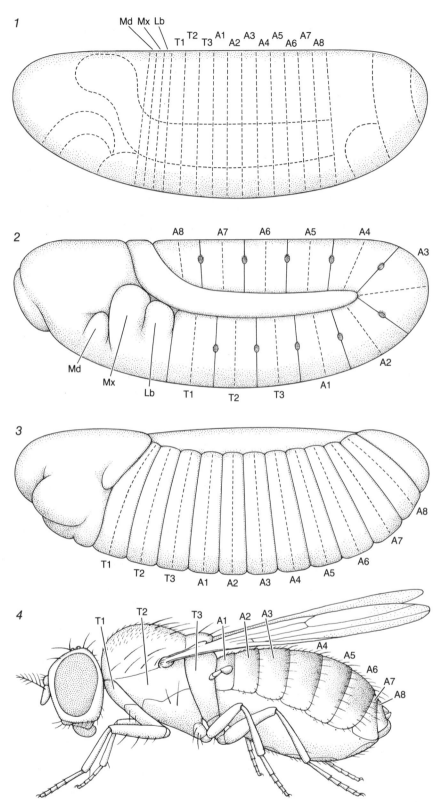

any part of the adult fly. The clone, or group of descendant cells, tend to be found together, but they can form a part of any adult structure.

Soon after the blastoderm is formed, however, the process of determination begins to move forward quickly. The initial determination affects both the precursors of the embryonic and larval segments and the precursors of the imaginal disks, which will form adult structures. Work done in my laboratory in the early 1970's showed that by the early blastoderm stage the imaginal-disk precursors become committed to forming either anterior adult structures or posterior ones. Early blastoderm embryos can be broken down into single cells and then "cultured" by transplanting the cells into the abdominal cavity of an adult fly or a larva. Cells treated in this way can undergo their complete development in culture and form adult structures. When the procedure was carried out, cells from the anterior part of the blastoderm gave rise only to anterior adult structures, whereas cells from the posterior part of the blastoderm gave rise only to posterior adult structures.

The fate of single cells was analyzed by marking individual cells in the blastoderm genetically and following their development to the adult stage, where the progeny of the marked cell form a clone. It was found that the clones respect segmental boundaries: even very large marked clones do not cross the boundary between one segment and the next. This result indicates that the cells of the blastoderm are already committed to becoming part of a particular segment. Additional information was gained by destroying small groups of blastoderm cells with a laser microbeam: a precisely focused beam of ultraviolet radiation. The fate of the destroyed cells can be deduced by observing the defects that appear in the adult. When the microbeam data are plotted on an image of the blastoderm, the result is a blastoderm "fate map," correlating blastodermal regions with body structures. The larval and adult segments appear on the fate map as an orderly pattern of stripes.

Thus the segmental architecture of *Drosophila* is already established at the blastoderm stage and groups of cells in the blastoderm have been assigned to one segment in the larval and adult body. The assignment of cell fates, however, is by no means complete at the blastoderm stage. Immediately after the segments are laid down each cell in the segment is assigned either to the anterior half of the segment or to the posterior half and the segment is thereby divided into two compart-

**LATER STAGES** of embryonic development in *Drosophila* include the division of the body into segments and complex morphogenetic movements. Soon after the blastoderm forms, regions of it become committed to forming specific body structures. By means of various experiments the structures have been traced back to the portions of the blastoderm where they originated. The result is a blastoderm "fate map," on which the body segments appear as consecutive stripes (*1*). The embryo elongates in a way that causes the abdominal segments to be displaced over the dorsal surface in the anterior direction (*2*). At this stage the *engrailed* gene is expressed and the segments are subdivided. The embryo shortens again, the abdominal segments move back toward the posterior and the segmental boundaries become visible (*3*). The embryo hatches into a larva that molts twice, pupates and becomes an adult (*4*). The adult is segmented, but its outer covering is not derived from that of the larva. Instead the adult structures stem from pouches called imaginal disks in the body of the larva.

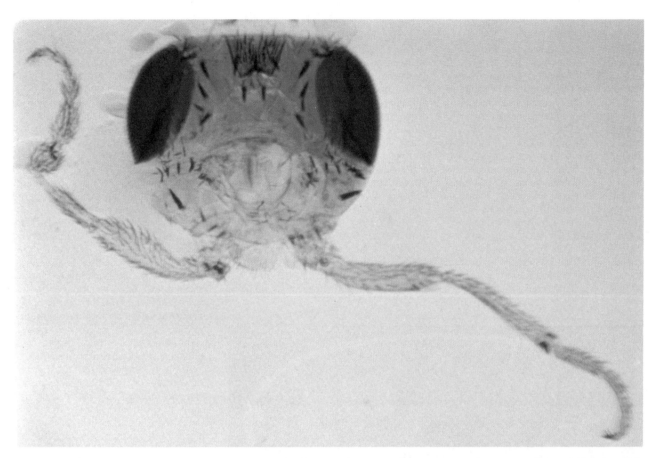

**NASOBEMIA is a mutation that causes legs to grow on the head of** *Drosophila* **in place of antennae. Apparently the mutation transforms the imaginal disks that would normally yield the antennae into the type of disk that yields the middle pair of the fly's three pairs of legs. From the effect of the mutation it was deduced that** **the effect of the normal gene is to ensure that each imaginal disk becomes committed to forming the correct adult structure.** *Nasobemia* **was discovered by the author in 1965. It is among a group of mutations that transform one structure into another normally found on a different segment; such mutations are called homeotic.**

ments: an anterior one and a posterior one. Following compartment formation the adult structures are determined. First all the imaginal disks are distinguished from one another. The disks contain many small regions, each one corresponding to a part of an adult structure such as the second segment of a leg. The small regions are assigned their destinies in a series of steps that apparently does not end until the final days of the larva. At metamorphosis each small group of determined cells in the imaginal disk is transformed under the influence of hormones into its adult counterpart.

The general conclusion of the genetic-marking work was that the cells in the *Drosophila* embryo are determined in a series of progressively finer gradations that terminates only at metamorphosis. How is this series of steps accomplished? Long before the era of molecular genetics some clues to the mechanism of determination were obtained from observations of three intriguing kinds of mutations that disrupt the developmental process in *Drosophila:* maternal-effect mutations, segmentation mutations and homeotic mutations. Certain maternal-effect mutations influence the spatial polarity of the embryo. For example, in a normal follicle the nurse cells are found only near the anterior pole of the egg. In the mutation known as *dicephalic* nurse cells are found at both poles. Such bipolar follicles give rise to embryos that have two sets of anterior structures joined in the middle and lack posterior structures altogether.

That the two-headed monsters are due to defects in the genome of their mothers can be concluded from genetic data. *Drosophila* females homozygous for the *dicephalic* mutation, which have two copies of the defective gene, produce aberrant follicles (and hence aberrant embryos) regardless of the genetic contribution of the father. Heterozygous mothers, which have one defective gene and one normal gene, produce only normal eggs. This result shows that the anterior-posterior polarity is laid down when the egg is formed in the ovary under the control of the maternal genome. Other maternal-effect mutants affect the dorsoventral polarity. Observations of maternal-effect mutants suggest the egg cytoplasm contains substances that define the spatial coordinates of the future embryo. After fertilization, when the nuclei migrate to the egg cortex, they encounter these substances and become committed to particular fates according to their position in the cortical cytoplasm.

The maternal-effect mutants are of interest because they suggest that some of the first steps in the process of determination are actually carried out under the influence of the maternal genome rather than the genome of the fertilized egg. Little is currently known, however, about the substances coded for by the maternal genes that endow the egg cytoplasm with its spatial polarity. There is some evidence that maternal messenger RNA (mRNA) stored in the egg has a role in specifying the dorsoventral polarity, but most of the other cytoplasmic substances remain mysterious. In contrast, the past few years have seen remarkable advances in the understanding of how segmentation genes and homeotic genes operate. Indeed, those two kinds of genes have provided the point of entry to the molecular level in developmental biology.

Most segmentation mutations and homeotic mutations are expressed only after the zygotic genome is activated during the formation of the blastoderm. Each segmentation mutation interferes in a particular way with the orderly division of the embryo into repetitive subunits. One of the most striking segmentation mutants is *fushi tarazu,* which means "not enough segments" in Japanese. In a *fushi tarazu* embryo parts of certain segments are missing and the incomplete portions are fused with adjacent segments. Thus the anterior portion of segments Mx, T1, T3, A2, A4, A6 and A8 are fused with the posterior portion of the segments behind them [*see illustration below*]. The result is an embryo

that has seven segments instead of 14 and dies before hatching into a larva. *Fushi tarazu* is one of a large group of mutations that affect the pattern of segments.

The most dramatic derangements of *Drosophila* development, however, are those produced by homeotic mutations. A homeotic mutation entails the transformation of one body part into another part that is normally found on a different segment. The results of such transformations are grotesque and intriguing. Wings grow where eyes belong, legs grow where the proboscis (feeding tube) belongs, legs become antennae, and so on. I have been fascinated by homeotic mutations since 1965, when, as a graduate student at the University of Zurich, I discovered a mutant that had legs on its head in

place of antennae. In seeking a name for the mutation I was reminded of a lyric by the German poet Christian Morgenstern describing a fantastic creature called the Nasobem, which walks on its nose. Amused at the correspondence, I decided to name the mutation *Nasobemia.*

The class of mutations to which *Nasobemia* belongs is large and diverse and has been the subject of considerable genetic work. It has been found that in *Drosophila* most of the homeotic genes are found in two clusters. One cluster, called the *Antennapedia* complex, consists of the genes that determine adult structures of the head and the anterior thoracic segments; *Nasobemia* is found in the *Antennapedia* complex. The other complex is called *bithorax* and includes the genes that control the determination of the posterior thoracic and abdominal segments. On the basis of his extensive analysis of the *bithorax* complex Edward B. Lewis of the California Institute of Technology proposed that each posterior segment of the adult is determined by the combined activity of a unique group of homeotic genes. In Lewis' model the determination of the second thoracic segment (the most anterior segment controlled by the *bithorax* complex) requires the fewest homeotic genes. Each successive segment posterior to that one requires the activation of one or more additional homeotic genes to take on its specific character.

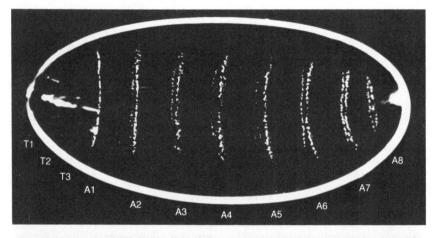

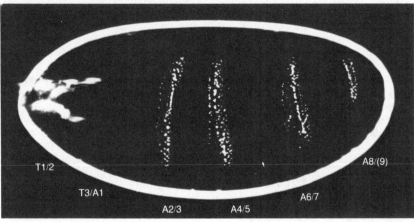

**FUSHI TARAZU is a segmentation gene: one whose action is needed for the *Drosophila* embryo to be divided into segments correctly. An embryo with the wild-type, or normal, *fushi tarazu* gene has the full complement of at least three head segments, three thoracic segments and eight complete abdominal segments (*upper panel*). In the process of development the head segments are retracted into the body and so are not visible. The anterior edge of each segment is marked by a belt of tiny projections called denticles. The denticle belts appear in the illustration as white stripes across the images. Embryos carrying a mutation in the *fushi tarazu* gene lack portions of alternate body segments and the remaining portions are fused (*lower panel*). For example, the posterior part of A2 is absent along with the anterior part of A3; the missing part of A3 includes the denticle belt. The remaining parts of A2 and A3 have fused to yield the compound segment A2/3. The mutation is lethal: it results in an embryo with half the normal number of segments that dies before hatching.**

The disruptive effect of the homeotic mutations and segmentation mutations was so profound that it suggested the corresponding genes served to orchestrate the normal process of development. Since a segmentation mutation interferes with the laying down of the embryonic segments, the normal gene is thought to regulate the correct construction of those segments. Since a homeotic mutation causes legs to grow where antennae should be, the normal gene ought to be responsible for the formation of legs in the proper place. Now, any gene capable of organizing such fundamental processes in normal development must work by regulating many other genes. Work in bacterial genetics had demonstrated that, by encoding a protein that can bind to DNA and inhibit or activate transcription, single genes can regulate the action of large groups of other genes. The bacterial model appeared to provide a plausible hypothesis for how the master genes of development might work. Yet as recently as 1978 such ideas were purely speculative because there were no convenient methods for isolating and working with developmental genes in higher organisms.

Within a few years the situation changed dramatically as gene-cloning methods made it possible to isolate genes in the absence of any biochemical information about their products. Immediately several research groups set out to purify homeotic genes from *Drosophila*. David S. Hogness of the Stanford University School of Medicine and his co-workers did pioneering work on the *bithorax* complex, while my group concentrated on the genes of the *Antennapedia* cluster. When the homeotic genes had been purified, they turned out to be surprisingly large and complex. For example, the *Antennapedia* gene spans 100,000 nucleotide base pairs, an unusually high figure. Moreover, the homeotic genes have a complex structure that includes many exons separated by introns. Introns are DNA sequences that are transcribed into RNA but subsequently removed from the transcript; only the exons are retained and spliced into the mature mRNA. In the processing of the *Antennapedia* transcript as many as 60,000 base pairs can be discarded in the form of a single intron.

Such understanding of the structure of the homeotic genes could not be achieved until some of them had been isolated. The *Antennapedia* gene, which gave its name to the *Antennapedia* complex, was isolated by Richard Garber and Atsushi Kuroiwa of my group. The first step was to assemble a set of consecutive, slightly overlapping DNA fragments spanning the chromosomal region where *Antennapedia* was known to be. This was accomplished by a method called "walking along the chromosome," which was originated by Hogness. Chromosomal walking relies on the fact that the two strands of the DNA helix are complementary and therefore can hybridize, or form a double-strand molecule. If two pieces of DNA from opposite strands of the double helix have a slight overlap, they will hybridize in the overlapping region. In walking one begins with a short DNA fragment that is known to be fairly near the desired gene. By means of hybridization a short piece of DNA overlapping the known fragment at one end and extending toward the desired gene is identified. By repeating the process the entire chromosomal segment including the gene can be covered.

After assembling the set of DNA pieces, it was necessary to find the precise location of the *Antennapedia* gene in the chromosomal segment. To find the gene we first isolated a cDNA

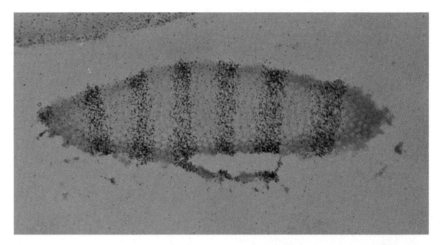

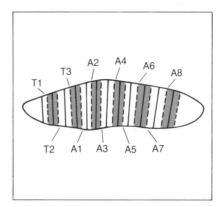

**FUSHI TARAZU IS EXPRESSED in pairs of segments in the early embryo. The illustration shows a thin section through an embryo two and a half hours after fertilization. The pale circles most clearly visible at the right end of the image are nuclei, which have just reached the cortex and are not yet separated by cell membranes. The dark bands are mRNA's from the *fushi tarazu* gene that have been located by in situ hybridization. (The exposed silver grains can be made dark or light in the photographic image by choosing the appropriate optics.) Each band spans two embryonic segments, as the key at the left shows. The section shown does not include the head segments. If it did, there would be seven dark bands instead of six.**

| | 1 | | | | | | | | | | | | | | | | | | | 20 |
|---|---|---|---|---|---|---|---|---|---|---|---|---|---|---|---|---|---|---|---|---|
| MOUSE *MO*-10 | Ser | Lys | Arg | Gly | Arg | Thr | Ala | Tyr | Thr | Arg | Pro | Gln | Leu | Val | Glu | Leu | Glu | Lys | Glu | Phe |
| FROG *MM*3 | Arg | Lys | Arg | Gly | Arg | Gln | Thr | Tyr | Thr | Arg | Tyr | Gln | Thr | Leu | Glu | Leu | Glu | Lys | Glu | Phe |
| *ANTENNAPEDIA* | Arg | Lys | Arg | Gly | Arg | Gln | Thr | Tyr | Thr | Arg | Tyr | Gln | Thr | Leu | Glu | Leu | Glu | Lys | Glu | Phe |
| *FUSHI TARAZU* | Ser | Lys | Arg | Thr | Arg | Gln | Thr | Tyr | Thr | Arg | Tyr | Gln | Thr | Leu | Glu | Leu | Glu | Lys | Glu | Phe |
| *ULTRABITHORAX* | Arg | Arg | Arg | Gly | Arg | Gln | Thr | Tyr | Thr | Arg | Tyr | Gln | Thr | Leu | Glu | Leu | Glu | Lys | Glu | Phe |

| | 21 | | | | | | | | | | | | | | | | | | | 40 |
|---|---|---|---|---|---|---|---|---|---|---|---|---|---|---|---|---|---|---|---|---|
| MOUSE *MO*-10 | His | Phe | Asn | Arg | Tyr | Leu | Met | Arg | Pro | Arg | Arg | Val | Glu | Met | Ala | Asn | Leu | Leu | Asn | Leu |
| FROG *MM*3 | His | Phe | Asn | Arg | Tyr | Leu | Thr | Arg | Arg | Arg | Arg | Ile | Glu | Ile | Ala | His | Val | Leu | Cys | Leu |
| *ANTENNAPEDIA* | His | Phe | Asn | Arg | Tyr | Leu | Thr | Arg | Arg | Arg | Arg | Ile | Glu | Ile | Ala | His | Ala | Leu | Cys | Leu |
| *FUSHI TARAZU* | His | Phe | Asn | Arg | Tyr | Ile | Thr | Arg | Arg | Arg | Arg | Ile | Asp | Ile | Ala | Asn | Ala | Leu | Ser | Leu |
| *ULTRABITHORAX* | His | Thr | Asn | His | Tyr | Leu | Thr | Arg | Arg | Arg | Arg | Ile | Glu | Met | Ala | Tyr | Ala | Leu | Cys | Leu |

| | 41 | | | | | | | | | | | | | | | | | | | 60 |
|---|---|---|---|---|---|---|---|---|---|---|---|---|---|---|---|---|---|---|---|---|
| MOUSE *MO*-10 | Thr | Glu | Arg | Gln | Ile | Lys | Ile | Trp | Phe | Gln | Asn | Arg | Arg | Met | Lys | Tyr | Lys | Lys | Asp | Gln |
| FROG *MM*3 | Thr | Glu | Arg | Gln | Ile | Lys | Ile | Trp | Phe | Gln | Asn | Arg | Arg | Met | Lys | Trp | Lys | Lys | Glu | Asn |
| *ANTENNAPEDIA* | Thr | Glu | Arg | Gln | Ile | Lys | Ile | Trp | Phe | Gln | Asn | Arg | Arg | Met | Lys | Trp | Lys | Lys | Glu | Asn |
| *FUSHI TARAZU* | Ser | Glu | Arg | Gln | Ile | Lys | Ile | Trp | Phe | Gln | Asn | Arg | Arg | Met | Lys | Ser | Lys | Lys | Asp | Arg |
| *ULTRABITHORAX* | Thr | Glu | Arg | Gln | Ile | Lys | Ile | Trp | Phe | Gln | Asn | Arg | Arg | Met | Lys | Leu | Lys | Lys | Glu | Ile |

**HOMEO DOMAIN is the string of amino acids corresponding to the homeobox. The homeobox is a short piece of DNA found in more than a dozen homeotic genes and segmentation genes from *Drosophila* and also in genes from a wide range of other higher organisms. The illustration shows the sequence of 60 amino acids (indicated by their three-letter codes) in the homeo domains arising from five genes. They are the mouse gene *MO*-10, the frog gene *MM*3 and three *Drosophila* genes: *Antennapedia*, *fushi tarazu* and *Ultrabithorax*. The *Antennapedia* gene has been employed as the standard for comparison. Discrepancies between the homeo domain of *Antennapedia* and those of the other genes are shown in white. The five homeo domains are quite similar. The resemblance among them suggests that the homeo domain serves the same function in all five proteins and that selective pressure has served to keep the amino acid sequences from varying much. It is not known precisely how the proteins that contain the homeo domain function. One clue is that all the homeo domains are rich in the basic amino acids lysine (*Lys*) and arginine (*Arg*), which could enable the homeo domain to bind to DNA. By binding to specific DNA sequences, these proteins could regulate the action of many other genes and so exercise a controlling influence on development. A stretch of nine amino acids that are identical in all five homeo domains is thought to be the region where the protein contacts the DNA (dark color).**

**DEFORMED is the name of a homeotic gene that appears to specify the identity of posterior head segments in the *Drosophila* embryo. The photograph shows a section through an embryo soon after the formation of the blastoderm; anterior is at the upper left, posterior at the lower right. The bright band gives the position of the *Deformed* transcripts as detected by in situ hybridization. On the fate map the place where the bright band appears corresponds to the posterior head segments. Furthermore, flies carrying a mutation in the *Deformed* gene have defects in those segments. It seems that the wild-type *Deformed* gene is needed for the posterior head region to be formed correctly. Considerable recent evidence suggests that the identity of *Drosophila* body segments is specified by homeotic genes.**

clone, a fragment of DNA complementary to the *Antennapedia* mRNA. The clone will hybridize to chromosomal DNA sequences that are complementary to the mRNA molecule. This clone was employed as a probe to find the location of the DNA sequence that encodes the *Antennapedia* mRNA. We were astonished to find our probe hybridized not only with the *Antennapedia* coding sequences but also with sequences in a neighboring gene. It turned out that the neighbor is *fushi tarazu* and so it became clear that *fushi tarazu* and *Antennapedia* share a short stretch of DNA. (The same observation was made independently by Matthew P. Scott of the University of Colorado at Boulder.) To test the possibility that the common sequence might be characteristic of homeotic genes we examined a gene in the *bithorax* complex called *Ultrabithorax*. To our pleasant surprise my colleague William J. McGinnis (now of Yale University) found the short stretch of DNA is also included in *Ultrabithorax*.

The startling discovery of a common DNA segment in my laboratory in 1983 initiated a rapid search through the *Drosophila* genome. Employing the short common sequence as a probe, we quickly identified more than a dozen genes containing similar sequences. Since many of the newly

isolated genes corresponded to known homeotic mutants, we designated the common sequence the homeobox. Indeed, it seems all the genes that include the homeobox are either homeotic or otherwise implicated in determining the spatial organization of the embryo. Most of them are in the *Antennapedia* complex or the *bithorax* complex. The genes of these two complexes were known to be connected with the determination of spatial organization and most of them had already been identified. Some of the other homeobox-containing genes, however, are in distant parts of the genome and have not been identified by means of mutations.

It is notable that among the genes containing the homeobox are segmentation genes such as *fushi tarazu* and genes known to be involved in the division of the segments into compartments, such as the one called *engrailed*. Although the physical effects of the segmentation genes and compartmentation genes may be quite different from those of the homeotic genes, the discovery of the homeobox in all three groups shows they have a significant common element. The discovery of that common element strengthens the hypothesis that all three kinds are among the master genes that orchestrate development. Whether all the homeobox-containing genes derive from

a common ancestor or resemble each other only in the exon that contains the homeobox is not yet known. Further work should soon resolve this significant question.

Having established the existence of the homeobox in the fruit fly, we looked for new genomes to probe. There was little surprise in finding the common sequence in other species of *Drosophila* and in insect species known to produce homeotic mutants, such as beetles. Examination of the annelid worms, the ancestors of insects, showed they also have the homeobox. That finding was intriguing. Quite another level of surprise, however, was provided by the finding that vertebrates share the homeobox. In collaboration with my colleague Edward De Robertis and his group, my group succeeded in isolating the first vertebrate homeobox from the frog *Xenopus laevis*. Subsequently all other vertebrates examined, including human beings, were found to have homologous, or similar, sequences. A brief preliminary analysis indicates that the homeobox is present in all groups of segmented animals. Although it has not been detected in most groups of animals that do not have segments, it does seem to be present in sea urchins.

The first clues to how the homeobox might operate at the molecular level came from comparing the DNA sequences of various homeoboxes. More than a dozen such sequences have now been determined. By comparing them it has been found that the homology among homeobox sequences is confined to about 180 base pairs. The homology ranges from 60 to 80 percent, depending on the genes compared. All the homeoboxes studied so far can be translated into an amino acid chain, which suggests that the homeobox codes for a domain, or functional segment, of a protein. This domain has been designated the homeo domain. The amino acid sequences corresponding to the homeoboxes are even more alike than the nucleotide sequences of homeoboxes themselves. (This is possible because a single amino acid unit can be coded for by more than one triplet, or group of three nucleotides.) For example, the amino acid chains corresponding to the homeoboxes of the *Antennapedia* gene and the *Xenopus* gene *MM*3 share 59 of their 60 amino acids, a remarkable overlap considering that vertebrates and invertebrates diverged more than 500 million years ago. The similarity among amino acid chains implies that all the homeo domains function in much the same way and that intense selective pressure has been exerted to prevent the function from being lost.

What might that common function be? As I noted above, an early speculation was that the master genes of development might operate by making proteins that bind to DNA. The homeo domain is biochemically compatible with such a model. Many DNA-binding proteins have regions that are rich in basic amino acids. The homeo domain, which forms a part of larger proteins, is rich in the basic amino acids lysine and arginine. A first indication that the homeo domain may be involved in DNA binding came from a computer search of known DNA sequences. The search detected a small but significant homology between the homeobox and portions of genes in two yeast species that are designated MAT genes. Each MAT gene codes for a protein that regulates all the genes needed to control the differentiation of the yeasts into one of two mating types or to form spores. The protein carries out its task by binding to specific DNA sequences that are "upstream" of the genes to be regulated (toward the 5′ end of the DNA). The partial homology between the homeobox and the yeast MAT-gene sequences suggests the homeo domain does the same.

If the homeo domain controls determination, the expression of the homeobox-containing genes must be precisely controlled temporally and spatially during embryogenesis. The spatial pattern of gene expression is currently being studied by in situ hybridization (in situ means in the tissues of the organism rather than in the test tube). That technique relies on the hybridization of a single-strand piece of DNA from a gene to the corresponding mRNA transcript. Since

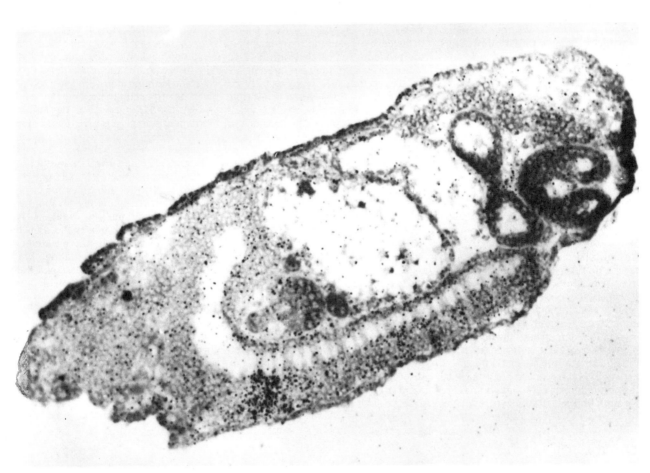

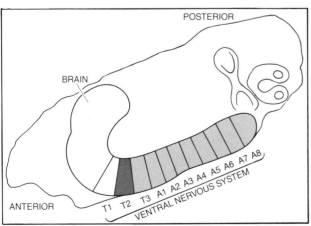

**ANTENNAPEDIA TRANSCRIPTS accumulate differentially in the segments of the embryo, as is shown in a section from an embryo in the later stages of embryonic development. In the later stages of embryogenesis segmentation is seen most clearly in the ventral nervous system, which forms one ganglion per body segment (key at left). *Antennapedia* is expressed very little in the first thoracic ganglion. It is expressed most strongly in the second thoracic ganglion and weakly in the segments posterior to that one. Differential expression may be part of the mechanism whereby homeotic genes bestow specific identities on the segments of the *Drosophila* embryo.**

the mRNA accumulates at the sites in the embryo where the gene is expressed, the hybridization can show the spatial pattern of gene expression. Probes are constructed by purifying homeobox-containing genes and labeling them radioactively. A drop of solution containing the probe is added to a thin section of a *Drosophila* embryo. The probe is allowed to hybridize to its mRNA transcript and the section is covered with a photographic emulsion. When the emulsion is exposed, the radioactive DNA probes bound to their own mRNA's appear as dark or light grains in the image depending on the optical system that is chosen. Ernst Hafen and Michael Levine refined the technique in my laboratory to make possible the detection of transcripts from homeotic genes.

Some of the most interesting results of in situ hybridization have come from using *fushi tarazu* as a probe. The *fushi tarazu* transcripts are first detected in the nuclei lined up in the cortical cytoplasm before cell membranes are formed. As the nuclei divide, the rate of transcription increases and soon the dark grains form a dramatic pattern of seven belts around the blastoderm. The fate map shows those belts correspond precisely to the seven sections missing in the *fushi tarazu* mutant. When the embryonic segments are formed, the transcripts are no longer detected. From such results one can infer two significant points about how *fushi tarazu* operates. Early in embryogenesis the normal gene must be expressed in alternate sections for the segmental plan to be laid down correctly; thereafter it is dispensable.

The fact that the normal *fushi tarazu* gene is expressed in a spatially precise fashion before the cell membranes are formed implies that the bare nuclei have a "sensor" enabling them to identify their position in the cortical cytoplasm. I propose that the sensor is an upstream control region in *fushi tarazu* and other genes involved in the elaboration of the segmental pattern. That hypothesis has been tested by constructing an artificial gene that includes the *fushi tarazu* upstream control region and a bacterial gene for an enzyme called beta-galactosidase. The expression of the synthetic gene is controlled by the *fushi tarazu* sequence, but the protein made by the gene is the bacterial one.

The artificial gene is inserted into a *Drosophila* embryo and the expression of the beta-galactosidase gene is detected by a staining reaction. When the procedure is carried out, beta-galactosidase is found in a pattern of seven stripes that matches precisely the pattern of the *fushi tarazu* transcripts. It is clear that after the nucleus enters the cortical cytoplasm, a substance interacts with the control region of *fushi tarazu* and turns that gene on or off according to the position of the nucleus in the cortex. The protein product of *fushi tarazu* in turn may go on to regulate a group of other genes in a precisely orchestrated pattern.

*Fushi tarazu* is only one of a series of genes whose expression has been examined by in situ hybridization. Other experiments have thrown light on how the segmental plan is filled in. The gene *engrailed* is known to be required in the posterior compartment of each segment. In situ hybridization indicates that soon after *fushi tarazu* is expressed the *engrailed* transcripts accumulate in a pattern of 14 narrow stripes corresponding to the posterior compartments of the segments. Apparently the embryo is first divided into segments, which are then subdivided to form compartments. The action of the homeotic genes can best be traced in the ventral nervous system, which later in embryonic development forms one ganglion per body segment. Employing the homeobox as a probe, it has been possible to clone a series of homeotic genes whose transcripts accumulate in consecutive ventral ganglia. Each gene is expressed most strongly in a particular segment and less strongly in all the segments posterior to that one. This is consistent with Lewis' hypothesis that the activity of a particular combination of homeotic genes specifies the identity of individual segments.

Intense work is now under way in many laboratories to consolidate and extend the knowledge gained by the identification of the homeobox. If the homeo domain does bind to DNA, it will be important to find out where and how. In addition, even if the homeobox-containing genes regulate many other genes, they must be regulated themselves. Finding out how the regulators are regulated will be another significant accomplishment, and one that could lead to the identification of the factors in the egg cytoplasm that provide the positional information. All this feverish work marks the division between two eras. In the premolecular era much was learned about when the embryonic cells are assigned their fates. In the molecular era, which has just begun, it will be possible to find out how this is done. Although animals develop in very diverse ways, the discovery of the homeobox in a wide range of species suggests that the molecular mechanisms underlying development may be much more universal than was previously suspected.

# 11

# THE MOLECULAR BASIS OF EVOLUTION

# The Molecular Basis of Evolution

*The discovery that mutations accumulate at steady rates over time in the genes of all lineages of plants and animals has led to new insights into evolution at the molecular and the organismal levels*

by Allan C. Wilson

The molecules of life are now the chief source of new insights into the nature of the evolutionary process. For a century the main contributors to knowledge of evolution were biologists working at the level of the whole organism. Together with geologists they established that the millions of kinds of creatures living today descended from a few species that lived more than a billion years ago. They also recognized that biological evolution results from heritable change made possible by mutation and natural selection. Until recently, however, investigators could not probe evolution at its most basic level. They could not directly explore changes occurring in genes.

New techniques in biochemistry have made such investigation possible. In recent decades molecular biologists have been able to compare the genes of thousands of living species and a few extinct species. They have measured the extent of the differences among the genes and studied the nature of the differences. One major result of the analysis is the concept of the molecular clock. Because mutations change the DNA in all lineages of organisms at fairly steady rates over long periods of time, one can establish a clocklike relation between mutation and elapsed time. Investigators have calibrated the clock on the basis of a few precisely dated fossils that yield estimates of the elapsed times since particular groups of living species diverged from common ancestors. Molecular differences can then be used to estimate the dates of divergence for multitudes of other species. Evolutionary biology has begun to acquire a quantitative molecular foundation.

My discussion of the molecular basis of evolution rests on two assumptions: (1) that the heritable differences among organisms result from differences in their DNA's, and (2) that molecular evolutionists must not only measure differences in DNA but also explain the origin of the differences and their relation to organismal differences. In this article I shall describe some of the discoveries and concepts of molecular evolution, attempt to relate it to organismal evolution and then argue that molecular biology has introduced a new way of analyzing organismal evolution. In particular I maintain that pressure to evolve arises not only from external factors such as environmental change but also from the brain of mammals and birds: from the power to innovate.

Two critical elements of molecular evolution are point mutations (specifically, those occurring in the genes coding for proteins) and regulatory mutations. A point mutation is a single replacement of a DNA base. Such a mutation can affect the amino acid sequence of a protein. A regulatory mutation, on the other hand, is any change in a gene or in the vicinity of a gene that determines whether the gene is active or inactive. The investigation of point mutations has resulted in the conceptualization of the molecular clock and in the discovery of a kind of genetic change known as a neutral mutation: a mutation that is neither advantageous nor disadvantageous for an organism. Work with point mutations has also yielded many important insights into the branching of lineages of species. The inclusion of regulatory mutations has led to an even more thorough understanding of the link between molecular evolution and organismal evolution.

In examining point mutations molecular biologists ideally would like to compare DNA structures directly. Before such comparisons became possible, however, chemists had discovered how to compare the structure of proteins [*see illustration on pages 122 and 123*]. There is a simple relation between the sequence of amino acids in a protein and the sequence of bases in the gene that codes for the protein. Specifically, each replacement of an amino acid in a protein can be ascribed to a point mutation in a gene. Investigators have therefore gained insight into molecular evolution by comparing amino acid sequences.

During the course of comparative studies of protein structure, several workers began considering how the number of amino acid replacements might be related to the time that had elapsed since any two species of organisms had a common ancestor. By simply counting the replacements (and thus ignoring their nature and their lo-

LYSOZYME, the enzyme modeled in the computer image on the opposite page, serves as a measure of regulatory mutations: changes in DNA that determine whether genes are active or inactive. The image depicts a molecule of lysozyme with part of its substrate (*purple*), the substance on which it acts. The substrate is a sugar polymer found in the cell wall of bacteria. Lysozyme cleaves the polymer and so breaks down the bacterial cell wall. It has therefore been recruited as a major digestive enzyme in ruminant animals (such as cows and sheep) to retrieve nitrogen and phosphorus present in the bacteria in their stomach. (The bacteria function in the digestion of cellulose.) The stomach of most other mammals, on the other hand, contains only a low concentration of lysozyme. The difference in the concentration of the enzyme contrasts with its functional uniformity: lysozymes from all mammals function nearly identically. The difference in concentration is largely attributable to regulatory mutations, which are thought to play an important role in organismal evolution. The computer image was made by the Graphics Systems Research Group of IBM U.K. Limited. The white balls correspond to carbon atoms, the red balls to oxygen, the blue balls to nitrogen and the yellow balls to sulfur; the white sticks represent interatomic bonds.

cation in the protein structure) they discovered that proteins behave like approximate evolutionary clocks. A great deal of evidence points to the fact that amino acid replacements accumulate at fairly steady rates over long periods of evolutionary time. Techniques that allow direct comparison of genes confirm the hypothesis that the steady evolution of proteins is rooted in the steady evolution of DNA. In nuclear DNA and in the DNA of other cellular components (such as mitochondria and chloroplasts), for example, the average accumulation of base replacements is nearly as clocklike as the process of radioactive decay.

The molecular clock, however, does not tick at the same rate at every position along the DNA molecule. The rate of evolution at a site in DNA that directly affects the function of a protein is slow; it is faster at a position that does not affect such a function. In other words, evolutionary change at the molecular level is slow where there are strong functional constraints and faster where they are weak. The active sites of most enzymes, for instance, evolve slowly compared with many other parts of the enzyme structure. The structures of other proteins also

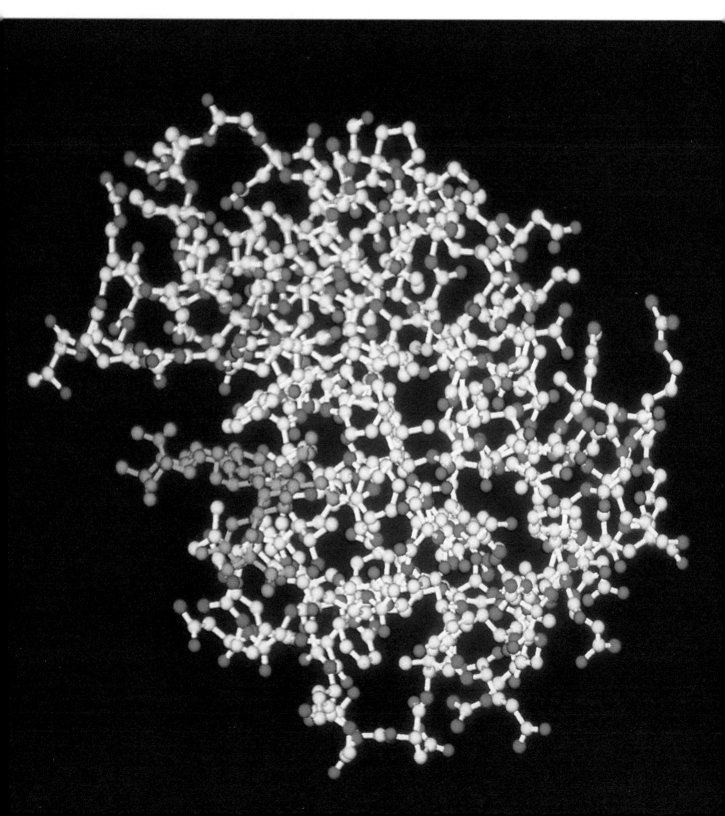

illustrate the concept of functional constraint. The hemoglobin of horses and that of human beings differ from each other by amino acid replacements at 43 out of 287 positions. In spite of these many differences, the chains of amino acids in these two hemoglobins are shown by crystallographic studies to fold in identical ways. Moreover, the two proteins behave nearly identically in functional tests: the point mutations ascribable to the 43 replacements are subject to weak functional constraints.

Comparisons of codons (triplets of DNA bases, each of which specifies a particular amino acid) provide a third example of functional constraint. The rate of change at the third position of codons is greater than the rate of change at the second position. This observation corresponds to the fact that whereas any base change at the sec-

ond position results in an amino acid substitution, about half of the base changes at the third position do not result in a substitution. The functional constraint on evolutionary change at the second position is strong because change occurring there is more likely to affect protein function; the constraint on change at the third position is weak because change can occur there without disrupting the function of proteins.

Observations of the high rate of evolutionary change at weakly constrained DNA positions have encouraged biologists to regard molecular evolution as an accumulation of neutral mutations that do not interfere with protein function. This way of looking at molecular evolution has been uncomfortable for Darwinists accustomed to thinking of evolution as resulting from the accumulation of ad-

vantageous mutations. The reconciliation of the two points of view lies in the fact that even though neutral mutations may dominate molecular evolution, the abundance of genetic variation allows for the accumulation of enough advantageous mutations to enable natural selection to have its effect at the organismal level.

The revolutionary idea that genetic change is dominated by neutral mutations has helped to explain the finding that molecular evolution depends more on years than on generations. If positive selection were driving molecular evolution, one would expect to find higher rates of evolution in short-lived species such as flies or mice than in long-lived species such as the higher primates. Instead base replacements accumulate at about the same rate in coding sequences along both kinds of lineages.

## PROTEINS

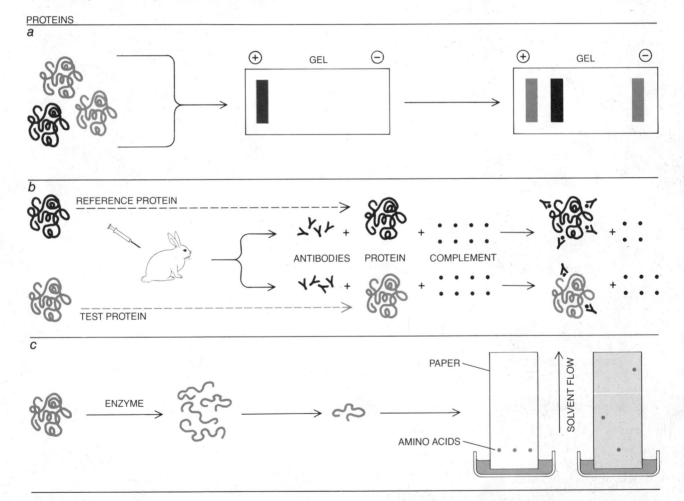

**MOLECULAR EVOLUTION** is measured by comparing proteins (*a–c*) or DNA's (*d–f*). Gel electrophoresis (*a*) can separate proteins on the basis of charge. Since the charge varies with the amino acid composition of a protein, the technique serves as a measure of the extent to which that composition varies in different versions of a protein. The method is most valuable when the electrophoretic mobilities of a set of 30 or more kinds of proteins from one individual are compared with the mobilities of the corresponding proteins from another individual. Microcomplement fixation (*b*) relies on the ability of antibodies to detect small differences between pro-

teins. Antibodies made by immunizing rabbits against a pure protein are tested in the presence of complement (a mixture of substances in blood) for their ability to bind with the immunizing protein and with related proteins. Complement interacts only with antibody bound to a protein antigen; the disappearance of complement measures the amount of antibody-antigen complex formed and therefore indicates differences in the proteins. In chemical sequencing (*c*) a purified protein is fragmented by an enzyme. The amino acids of each fragment are cleaved sequentially, beginning at one end of the fragment, and are identified by chromatography, a process in which

Nevertheless, many biologists who make mathematical models of the evolutionary process are coming to believe many of the mutations accumulated during molecular evolution are not neutral. They argue that instead of proceeding smoothly, molecular evolution might be characterized by long periods of inactivity punctuated by bursts of change. If they are right, the challenge of finding an explanation for the molecular-clock phenomenon grows. The explanation of the phenomenon may entail a deeper grasp of the nature of the evolutionary process.

On one point all molecular biologists agree: changes in the sequence of DNA's and the proteins they encode are mainly divergent. Investigators can therefore construct molecular trees, or branching diagrams, showing the genealogical relations among

these sequences. Such diagrams help one to think clearly and quantitatively about how present-day sequences evolved from a common ancestral sequence. Molecular trees also illuminate the genealogical pathway by which the species containing the sequences evolved from a common ancestral species. The order of lineage branching leading to modern species provides a valuable framework in which to organize knowledge of the differences among species.

To choose among alternative genealogical hypotheses molecular biologists follow the principle of Occam's razor: the simplest of competing theories is selected over more complex ones. The tree is chosen that requires the fewest mutations to explain the evolution of particular sequences from a common ancestral sequence. This approach allows molecular evolutionists

to choose objectively and quantitatively among alternative trees. How, for example, are human beings related to orangutans and African apes (chimpanzees and gorillas)? A branching diagram linking human beings most closely to African apes explains the molecular data by postulating fewer mutations than are required for diagrams linking humans most closely to orangutans. In other words, one diagram explains the observed sequence diversity so much more simply than others do that the complicated ones can be ruled out statistically.

The capacity to make such determinations is one of the notable achievements of molecular evolutionary biology. Previously investigators had based trees exclusively on differences in anatomical traits. The comparison of such traits is highly subjective. In addition the workers had no way of

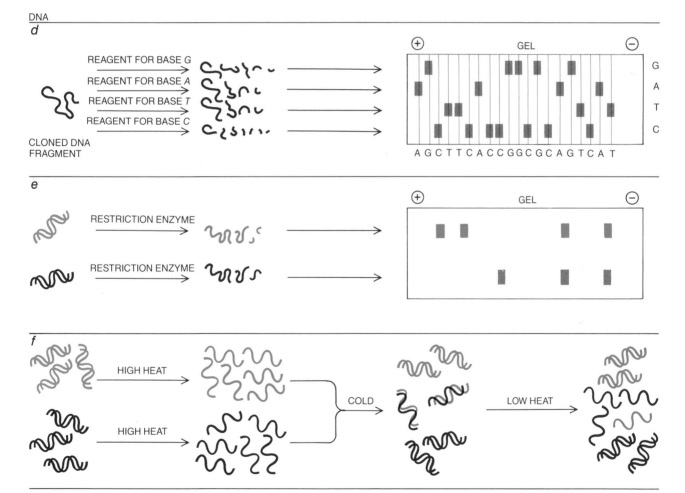

DNA

the migration of amino acids depends on size and charge. Two cloned DNA's can be compared in detail by sequencing (d). A piece of DNA, to which a radioactive label has been attached at one end, is cleaved by a reagent specific for one of the four DNA bases (G, A, T, C) under conditions such that each molecule is on the average cleaved at only one of the susceptible sites. The DNA sequence (AGCTTCACCGGCGCAGTCAT in this case) is inferred by reading the distances the cleaved fragments move through a gel under the influence of an electric field. A faster but less accurate method for comparing DNA's is restriction analysis (e). A piece of DNA is fragmented by a set of restriction enzymes, each of which recognizes and cleaves a specific sequence of from four to six bases. Differences in the sequences affect the size of the fragments, so that the pattern of fragments of two DNA's subject to cleavage and electrophoresis reflects their degree of similarity. In DNA hybridization (f) the double helix of analogous DNA's from two sources is disrupted by heating. When the two sets of single strands are cooled together, hybrid duplexes consisting of a strand from each of the DNA's can form. The stability of the hybrids to heat is a measure of the degree of sequence similarity between the DNA's.

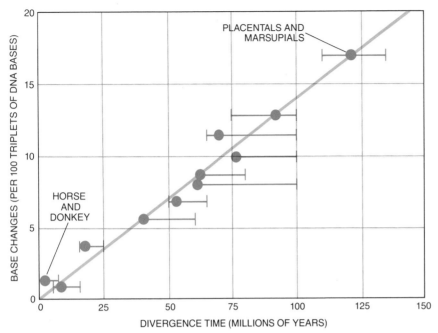

CLOCKLIKE EVOLUTION is shown for the genes of mammals whose times of divergence are known approximately from fossil evidence. The amino acid sequence for each of seven proteins was determined for 11 pairs of mammals or mammalian groups and the number of amino acid differences between the two members of each pair was calculated. The number of point mutations, or replacements of individual DNA bases, required to account for those differences was estimated, and it is indicated on the vertical axis of the graph. The horizontal axis indicates how long ago the particular lineages of each member of a pair diverged from each other. The most distantly related mammalian groups compared are placentals and marsupials, whose common ancestor lived about 120 million years ago. The most closely related pair are the horse and the donkey. The bars indicate the uncertainty in the estimates of divergence time. The curve shows that DNA-base replacements accumulate at fairly steady, or clocklike, rates over long periods of evolutionary time.

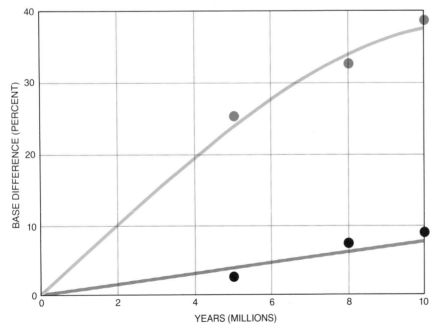

FUNCTIONAL CONSTRAINT is illustrated by comparing the rate of change at the second position of codons (black curve) to the rate of change at the third position (colored curve). A codon is a triplet of DNA bases that encodes a particular amino acid. Change takes place more rapidly at the third position than at the second one. The reason is that whereas any base change at the second position results in an amino acid substitution, about half of the base changes at the third position do not result in a substitution. The data are from comparisons made of the DNA of mitochondrial (cellular organelles) of apes and humans. In broader terms, the rate of evolution at a site in a gene that directly affects the function of a protein is slow; it is faster at a position that does not affect any such function.

knowing the number of mutations necessary to produce an observable difference in a trait. They also could not know whether a mutation giving rise to a difference in one anatomical trait also contributes to differences in other anatomical traits. Molecular trees built from sequence data require no subjective decisions about traits. Moreover, biologists know the minimum number of base replacements needed to account for the sequence differences. Finally, the number of countable genetic traits revealed by comparison of DNA and protein sequences has begun to exceed the number of anatomical traits available for tree analysis.

In addition to disclosing the order of lineage branching, molecular trees contain information about times of divergence among lineages. The first application of this approach to evolutionary dating involved estimating when hominoids such as human beings and African apes diverged from a common ancestor. Working in my laboratory at the University of California at Berkeley, Vincent M. Sarich measured the structural differences of serum albumin, a protein found in both humans and African apes. He determined the mean rate of evolutionary change by comparing the albumins of species whose divergence times were known from fossil evidence. He was thus able to calculate that humans and African apes diverged five million years ago. This was only a fraction of the time postulated by anthropologists: from 20 to 30 million years. Subsequent DNA studies have confirmed Sarich's work, leading to a reinterpretation of the fossil record and a revision in thinking about the pathway of evolution from ape to man.

Molecular trees have yielded many other insights into the genealogical links between species. Trees based on fast-evolving DNA positions link species that diverged rather recently (such as the hominoids). Whereas these positions facilitate the exploration of the twigs of the evolutionary tree, highly conserved positions allow the probing of the deepest branches. Genes containing many highly conserved positions reveal four primary branches of descent. The branches diverged from one another nearly three billion years ago, when all cells were at the bacterial level of organization. The pattern of branching offers new insights into the sequence of steps characterizing the evolution of metabolism in early cells.

Tree analysis has also supported the theory that eukaryotic cells (the nucleated cells of organisms higher than bacteria) arose by the fusion of two or more types of bacterial cells about a billion years ago. Eukaryotic cells contain DNA in distinct compartments: the nucleus, the mitochondrion and, in the case of photosynthetic cells, the chloroplast. The genome of each compartment includes a set of very conservative genes specifying the structure of RNA molecules in the ribosomes (the organelles on which proteins are assembled) of that compartment. Sequence comparisons show that whereas the ribosomal RNA genes in the nucleus stem from one of the four primary branches in the bacterial tree, those in the chloroplast and the mitochondrion stem from another.

Tree analysis has also helped to explain how the nuclear genome of eukaryotic cells has grown hundreds of times bigger than the bacterial genome. The pattern of genealogical relations among genes and other repetitive sequences within the nucleus offers clues about the steps involved in the process. These steps include the duplication of entire genes and their movement to new locations in the genome. The duplicate genes usually diverge independently, either acquiring new functions or becoming inactive pseudogenes: duplicate stretches of DNA that contain mutations preventing them from encoding a functional polypeptide, or short protein

chain. In other cases the duplicated DNA's communicate (exchange genetic information) with one another at varying rates as they evolve.

In addition tree analysis has contributed to knowledge of the evolutionary role of gene transfer between species that do not interbreed. Some viruses and plasmids (small circles of bacterial DNA) can transfer cellular genes from one species to another, but the stable integration of such genes from one species into the genome of another species is rare in nature. If it were common, the genome of each species would be a mosaic made up chiefly of horizontally transferred contributions from diverse species. In that case at-

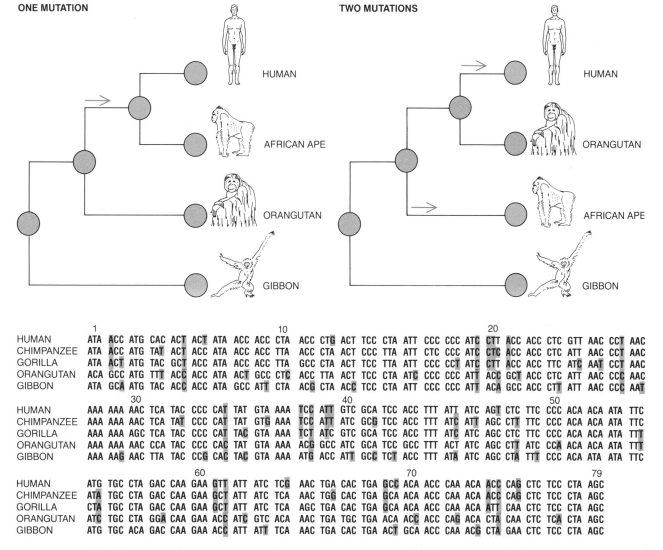

**BRANCHING DIAGRAMS,** or phylogenetic trees (*top*), showing the descent of the hominoids can be constructed on the basis of DNA sequences such as those at the bottom. Colored disks in the diagrams show the presence of a particular base (or amino acid) at a given position in a DNA (or protein) sequence for both humans and African apes (chimpanzees and gorillas). Gray disks show that a different base (or amino acid) is present at that position in orangutans and gibbons (Asian apes). The diagram at the upper left accounts for the sequence differences among the hominoid lineages with one mutation on the lineage leading to the common ancestor

of humans and African apes (*arrow*). The diagram at the upper right, in contrast, requires two mutations (*arrows*) to account for the data; it is less likely to be correct. The order in which humans diverged from chimpanzees and gorillas is still in dispute. The 79 codons shown in the bottom section of the illustration code, in the various hominoids, for part of a protein (NAD dehydrogenase 5) that functions in energy production within the mitochondrion. The sequences differ mostly by base replacements at third positions of codons (*gray panels*). The 16 colored panels indicate positions at which the African ape sequence resembles the human sequence.

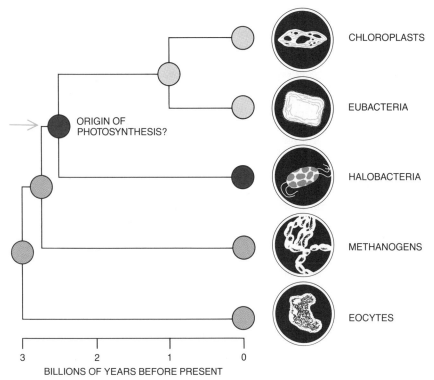

EVOLUTION OF EARLY CELLS led to the emergence of four main groups of bacteria: eubacteria (the major present form), halobacteria, methanogens and eocytes (sulfur bacteria). Chloroplasts (the photosynthetic organelles of eukaryotic cells), which arose from eubacteria about a billion years ago, share with them and halobacteria a capacity for photosynthesis. This capacity could have originated in the common ancestor of eubacteria and halobacteria. This phylogenetic tree was inferred from comparisons of ribosomal structure in the various organisms. Although there is uncertainty about the exact order of branching in the tree, it is widely agreed that the four primary branches are all of great antiquity.

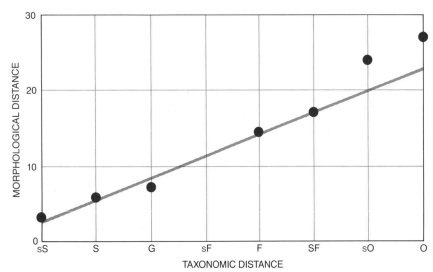

MORPHOLOGICAL DISTANCE, a measure of the extent to which animals differ in body plan, correlates with distance in taxonomic classification. The graph summarizes the results of more than 20,000 anatomical measurements on nearly 400 species and suggests that the accumulation of point mutations cannot explain the accelerated rate of organismal evolution in mammals. To estimate the morphological distance between two animals a standard set of bones from the head, forelimb, trunk and hindlimb of each animal is measured. The relative length of each trait is then calculated by dividing the length of that trait by the sum of the lengths of all traits measured in that animal. The morphological distance between two animals is the sum of the absolute values of the differences in the relative lengths for all traits. The smallest taxonomic distances in the classification are those between subspecies (sS), followed by those between species (S), genera (G), subfamilies (sF), families (F), superfamilies (SF) and suborders (sO). The largest taxonomic distances shown are for different orders (O). The points represent the mean morphological distances among birds. The line was fitted to analogous morphological-distance values for frogs, lizards and mammals.

tempts to build a tree for a set of species would prove futile; a tree based on one particular gene would probably disagree with a tree based on another gene. In practice, however, trees based on several different genes usually agree with one another. Most purported cases of horizontal transfer do not receive support from tree analysis. In both the bacterial and the eukaryotic worlds the predominant mode of evolution has been vertical: from parent to offspring.

Although the investigation of point mutations has increased understanding of evolutionary processes, it has failed to describe completely the link between molecular and organismal evolution. The sharp difference in the rates of organismal evolution for two groups of species, frogs and mammals (such as cats, bats, whales and humans), for instance, does not reflect the similarity in the rates at which point mutations accumulate for both groups. Frogs are an ancient group of animals consisting of thousands of species. Yet they share so many anatomical similarities that zoologists classify all frogs in one order. Indeed, during the period that saw the rise of cats, bats, whales and humans from a common ancestor, one type of frog evolved so slowly that both fossils 90 million years old and the present-day representatives of its lineage are classified in the same genus, *Xenopus*. Placental mammals, on the other hand, even though they represent a younger group, differ so much from one another that zoologists classify them in 16 orders.

Facts such as these indicate that the pace of organismal change in mammals has been much faster than it has been in frogs. Yet point mutations accumulate in the DNA of mammals at the same rate as they do in frogs. Similar contrasts between the rate at which point mutations accumulate and the rate of organismal evolution characterize many other groups.

The argument that there is a contrast between the rate of accumulation of point mutations and the rate of organismal evolution rests on the supposition that taxonomic classifications summarize, without bias, information about the degrees of anatomical similarity among species. To assess the validity of this assumption, Lorraine M. Cherry, then at the University of California at Berkeley, and Susan M. Case of Harvard University collaborated with me in developing a quantitative and objective way of estimating the degree to which species differ in body plan [*see illustration at bottom left*]. The results from our method agree with those obtained from traditional taxonomic methods.

The work of Cherry and Case lends quantitative support to the notion that the accumulation of point mutations cannot explain the accelerated rate of organismal evolution in mammals. The recognition of this discrepancy has led molecular biologists to ask two questions: What relation exists between molecular evolution and evolution at higher levels of organismal organization? What makes mammals evolve so fast at these higher levels?

One possible answer to the first question is that the majority of point mutations accumulating in nucleic acids and in the proteins they encode may be neutral or nearly so from the standpoint of natural selection. Only a minority may underlie adaptive evolution at the organismal level. The fraction of accumulated mutations having adaptive significance could be higher for mammals than for frogs but still too low to contribute significantly to the overall rate of molecular evolution in mammals.

In all likelihood, however, it is the regulatory mutation that establishes the link between molecular evolution and organismal evolution. A regulatory mutation is any mutation that affects the expression of a gene: particularly the turning on or off of specific genes in the course of development. In particular, attention has been paid to the idea that most adaptive evolution at the organismal level is due to mutations affecting the relative concentrations of specific proteins rather than to mutations affecting their structures.

To test these ideas one needs a strategy for picking genes with which to link molecular change to organismal change. Until the molecular basis of embryonic development is better understood, it does not seem profitable to search for those genes whose differences account for the anatomical differences between species of multicellular organisms. The best strategy at present is to work at the chemical interface between organism and environment. That is why investigators in my laboratory such as Deborah E. Dobson, Caro-Beth Stewart, R. Tyler White, Michael S. Hammer and Ellen M. Prager have probed genes coding for enzymes in the mammalian gut. The biochemistry and digestive physiology of mammals are well-developed subjects. Mammalian species diverge quickly from one another with respect to their diets. Biochemists can often guess which enzymes are necessary to cope with a chemical present in one diet but not in another. Genes coding for such enzymes therefore hold a key

GEOLOGICAL CHANGE

ENVIRONMENTAL CHANGE

BRAIN

CULTURAL CHANGE

NEW SELECTION PRESSURES

**PRESSURE TO EVOLVE, the author argues, comes both from geologic forces such as erosion and mountain building and from the brain of mammals and birds. By suddenly exploiting the environment in a new way, a relatively big-brained species quickly subjects itself to new selection pressures that foster the "fixation" of mutations complementary to a new habit. A mutation is said to be fixed in a population when descendants that bear the mutant gene predominate greatly over those individuals that bear the original gene.**

to understanding the relation between molecular and organismal evolution.

The investigation of bacteria-digesting enzymes confirms the significance of regulatory mutations. Although most mammals are not enzymically equipped to digest bacteria, on several occasions during mammalian evolution species have acquired the necessary enzymes. Ruminant animals such as cows and sheep, for example, need to digest bacteria in order to retrieve nitrogen and phosphorus that has been captured by the microorganisms. (The bacteria function in the digestion of cellulose.) The enhanced ability to digest bacteria is due to the presence of the enzyme lysozyme, which cuts open the cell wall of bacteria. Ruminant stomachs contain high concentrations of lysozyme, whereas most other mammalian stomachs contain low concentrations of the enzyme. Lysozyme has evidently been recruited as a major digestive enzyme in ruminants.

Although the recruitment of lysozyme depends on both regulatory mutations and structural mutations, regulatory change appears to have had the primary role. A similar picture emerges from studies of evolution in the test tube. The net conclusion from many experimental studies of evolution with both bacterial and animal cells in culture is that regulatory mutations may play a primary role in adaptive evolution.

The specific kind of regulatory mutation, however, remains unknown for many evolutionary processes. Although gene duplications and point mutations in regulatory DNA are responsible for most of the altered rates of protein synthesis observed in laboratory experiments, for instance, they may not account for the changing lysozyme levels in mammalian evolution. Because the lysozyme changes are tissue-specific, enhancers (regulatory DNA sequences recognized by factors specific to a given tissue) may prove to be responsible for controlling the levels of the enzyme. The tissue-specific activation of a gene can be accomplished by moving an enhancer into any one of a variety of noncoding positions within or near the gene. It remains to be seen whether the recruitment of lysozyme depends on enhancers and whether the lysozyme case typifies that of other genes taking part in major adaptive shifts.

The final question I address is why mammals evolve so fast at the organismal level. I maintain that the high rate of evolution for mammals with respect to that for frogs may be due to the large brain of mammals. A large brain generates an internal pressure to evolve that frogs lack. In reaching this conclusion I assume that organismal evolution is a Darwinian process driven by selection and therefore has two components: mutation and fixation. In other words, although a newly arisen mutation is initially present in a single individual within a population, the mutation has not been "fixed" until descendants bearing the mutant gene predominate greatly over individuals bearing the original type of gene. Quantitatively, the basic equation of evolution states that the rate of evolution within a population equals the number of mutations arising per unit of time multiplied by the fraction of those mutations destined to be fixed.

The high rate of mammalian evolution might therefore be attributed to either a large number of mutations or a large fraction of fixation, or to both. Even though the mammalian genome may indeed be more prone to mutation, or more unstable, than the genomes of "living fossils" (such as *Xenopus*) are, a large fraction of fixation seems more likely to account for the trend. In particular I consider the following possibility: the number of mutations arising per unit of time is the same for frogs and mammals, but the fraction of those mutations that are fixed is higher for mammals than it is for frogs. This would mean that mammals fix a larger fraction of their morphological mutations than frogs.

The opportunity to fix advantageous

mutations arises whenever the direction of selection changes. There are two basic sources for change in the direction of selection, which is to say there are two basic pressures to evolve adaptively. One comes from outside a species and the other comes from inside. Evolutionary biologists have tended to think only about the external factors, such as environmental change, which is largely driven by such geologic forces as erosion and mountain formation. The second pressure to evolve comes from the brain of mammals and birds. This internal pressure, a consequence of the power of the brain to innovate and imitate, leads to culturally driven evolution.

Once a species has a dual capacity to evolve, a new way of exploiting the environment can arise in a single individual and spread rapidly to other individuals by imitative learning. By suddenly exploiting the environment in a new way, a big-brained species quickly subjects itself to new selection pressures that foster the fixation of mutations complementary to the new habit. The larger the population is to which the new habit is communicated, the more likely it is that such a mutation will already be available, or will arise, so that selection can act on it. The time required for a population to fix a mutation that complements a new behavior is shorter if the new behavior spreads quickly not only to offspring (vertically) but also to other members of the population (horizontally).

The lineage leading to the human species has been under the highest internal pressure to evolve. The rise of agriculture, for instance, imposed new selection pressures that led to swift genetic changes in human populations. Consider the introduction of milk sugar (lactose) into the diet of adults as the result of the invention and social propagation of dairy farming. The genetic capacity of adults to digest this sugar has evolved only within populations dependent on dairy products. In the short period of 5,000 years genes conferring the ability to handle milk sugar as an adult reached a level of 90 percent in populations that depended heavily on dairy farming. In contrast, the level of the genes is virtually zero in human populations that do not drink milk and in all other mammalian species tested.

The potential for culturally driven evolution is by no means confined to humans. Imitative learning occurs in many species having brains that are relatively large in relation to body size, such as primates and songbirds. Imitative learning may also occur in some fishes, squids and insects, although it has not yet been demonstrated in them. The most celebrated case of a rapid shift in nonhuman behavior was provided by songbirds known as British tits. Some of these songbirds, which resemble American chickadees, learned how to open milk bottles. Soon they were imitated by millions of other tits. Within a couple of decades most of these British songbirds were engaging in the practice. Finally human beings stopped the evolutionary experiment: they put the bottles in crates. Biologists did not have the opportunity to ascertain whether or not the songbird population responded genetically to the new selection pressures generated by their new behavior.

My work with Jeff S. Wyles of Berkeley and Joseph G. Kunkel of the University of Massachusetts at Amherst supports the hypothesis that the brain of mammals and birds is the major driving force behind their organismal evolution. The investigators found that the larger the size of the brain in relation to the size of the body, the

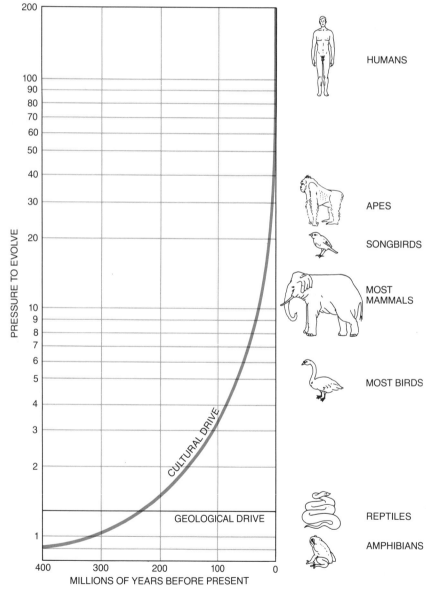

**INCREASE IN BRAIN SIZE** with respect to body size supports the theory that high selection pressure for mammals comes from the brain. A dramatic rise in the pressure to evolve at the organismal level is postulated to have occurred in the lineages leading from early amphibians to present-day species of birds and mammals. The vertical axis represents a measure of pressure to evolve. The curve for cultural drive was determined by dividing the mass of a species' brain (in grams) by the total mass of its body (in kilograms) raised to the two-thirds power. The level of the line plotting geologic drive is based on the assumption that pressure coming from geologic change has not undergone a net increase in the past 400 million years. Because humans, apes and songbirds have relatively large brains, they are under higher pressure to evolve than most mammals, birds, reptiles and amphibians.

**BRITISH TIT, a songbird, perches on a milk bottle after pecking open the foil cap. In the 1930's and 1940's the practice of opening milk bottles spread throughout the tit population of Britain, providing the most celebrated case of a cultural shift known in nonhumans. Human beings finally stopped the practice by putting the milk bottles in crates; biologists did not have an opportunity to learn whether or not the songbird population would have responded genetically to the new selection pressures that were generated by their behavior.**

but less pronounced tendency for the relative size of the brain to increase over time. In contrast, the relative brain size of modern frogs and salamanders does not differ significantly from the relative brain size of the first amphibians.

Since the rate of organismal evolution correlates with relative brain size, its rate may also have risen by a factor of nearly 100 along the lineage leading to humans and by smaller factors along the lineages leading to other big-brained creatures. Organismal evolution in the vertebrates may provide an example of an autocatalytic process mediated by the brain: the bigger the brain, the greater the power of the species to evolve biologically. When cultural evolution becomes extremely fast, however, such a process presumably does not keep accelerating. In such a case the pressures generated by one cultural shift will sometimes be relieved by the next cultural shift, rather than by a genetic response. This has probably been true of the human species for some 35,000 years, when the human brain reached its present size.

The study of molecular evolution occupies a special position in contemporary biology. In trying to link gene to organism, it touches molecular biology, cell biology, developmental biology, physiology, anatomy and behavioral biology. It also requires an understanding of how genes behave in populations, and the disciplines of taxonomy, paleontology and geology are involved. No other field touches all these aspects of biology and geology. The study of molecular evolution provides an opportunity to build bridges between biological disciplines and by so doing contribute to the unification of the life sciences.

higher the mean rate of anatomical evolution. During the evolution of vertebrates on land, the relative size of the brain has increased by a factor of 100 along the lineage leading from the first amphibians to humans. Furthermore, the rate of increase in relative size has accelerated. The lineages leading from those same early amphibians to other mammals and to birds exhibit a similar

# THE AUTHORS

# BIBLIOGRAPHIES

# INDEX

# THE AUTHORS

ROBERT A. WEINBERG ("The Molecules of Life") is professor of biology at the Center for Cancer Research of the Massachusetts Institute of Technology and a member of the Whitehead Institute for Biomedical Research. His B.A. (1964) and Ph.D. (1969) are both from M.I.T. He did postdoctoral research with Ernest Winocur at the Weizmann Institute of Science in Israel and with Renato Dulbecco at the Salk Institute for Biological Studies. In 1972 he returned to M.I.T., and the following year he was made a member of the faculty at the Center for Cancer Research. In 1982 Weinberg became professor and joined the Whitehead Institute.

GARY FELSENFELD ("DNA") is chief of the physical chemistry section in the Laboratory of Molecular Biology at the National Institute of Arthritis, Diabetes, and Digestive and Kidney Diseases. He received his A.B. in biochemical sciences at Harvard College in 1951 and studied physical chemistry with Linus Pauling at the California Institute of Technology, where he earned his Ph.D. in 1955. After three years of research at the National Institute of Mental Health he was appointed to the faculty of the University of Pittsburgh as assistant professor of biophysics. In 1961 he accepted his present position.

JAMES E. DARNELL, JR. ("RNA"), is Vincent Astor Professor at Rockefeller University. He got his B.A. at the University of Mississippi in 1951 and his M.D. at the Washington University School of Medicine in 1955. After medical school he was on the staff of the Laboratory of Cell Biology at the National Institutes of Health. In 1961 he moved to the Massachusetts Institute of Technology,

and in 1964 he joined the faculty of the Albert Einstein College of Medicine in New York. Beginning in 1968 Darnell spent six years at Columbia University, after which he took up his position at Rockefeller University.

RUSSELL F. DOOLITTLE ("Proteins") is professor of biochemistry at the University of California at San Diego. He has a B.A. from Wesleyan University (1952), an M.A. from Trinity College (1957) and a Ph.D. from Harvard University (1962). In 1964 he accepted a position as assistant research biologist at San Diego; three years later he was appointed assistant professor. He was made full professor in 1982. Doolittle traces his lifelong interest in the structure and evolution of proteins to a summer project comparing fish and human physiology as a graduate student.

MARK S. BRETSCHER ("The Molecules of the Cell Membrane") is head of the cell biology division of the Medical Research Council's Laboratory of Molecular Biology in Cambridge. He received his undergraduate and graduate training at the University of Cambridge, where he got a Ph.D. in 1964. After spending a year as a Fulbright scholar at Stanford he joined the scientific staff of the MRC Laboratory. In 1984 he was made head of its cell biology division. In July Bretscher finished a term as visiting professor of biochemistry at the Stanford University School of Medicine.

KLAUS WEBER and MARY OSBORN ("The Molecules of the Cell Matrix") are a husband-and-wife team investigating the cell matrix at the Max Planck Institute for Biophysical Chemistry in Göttingen. Weber, who is a director of the institute, earned his

Ph.D. in Germany before joining the faculty of Harvard University in 1967. After his return to Europe he served as secretary general of the European Molecular Biology Organization. Osborn is a member of the scientific staff at the institute. She did her undergraduate work at the University of Cambridge and got a Ph.D. in biophysics from Pennsylvania State University in 1967. She did postdoctoral research with James D. Watson at Harvard and then returned to her native England to work for three years at the Medical Research Council's Laboratory of Molecular Biology in Cambridge. After marrying in 1972 Weber and Osborn spent two years at the Cold Spring Harbor Laboratory before moving to Göttingen.

SUSUMU TONEGAWA ("The Molecules of the Immune System") is professor of biology at the Center for Cancer Research of the Massachusetts Institute of Technology. He was born in Japan and received his undergraduate education at Kyoto University. In 1963 he came to the U.S. for graduate study, and he got his Ph.D. from the University of California at San Diego in 1968. After a few more years in San Diego he moved to the Basel Institute of Immunology. He went to M.I.T. as professor in 1981. In 1983 Tonegawa was named Person with Cultural Merit by the Japanese government.

SOLOMON H. SNYDER ("The Molecular Basis of Communication between Cells") is director of the department of neuroscience at the Johns Hopkins University School of Medicine; he is also Distinguished Service Professor of neuroscience, pharmacology and psychiatry. He did his undergraduate work at Georgetown University and earned his medical degree from the Georgetown Medical School in 1962. In 1965 he went to the Johns Hopkins Hospital as an assistant resident in the psychiatry department. A year later he joined the faculty at Johns Hopkins as assistant professor of pharmacology and experimental therapeutics. He was made professor in 1970.

MICHAEL J. BERRIDGE ("The Molecular Basis of Communication within the Cell") is senior principal scientific officer in the unit of insect neurophysiology and pharmacology at the University of Cambridge. He was born in Zimbabwe and received a B.Sc. degree with first-class honors from the University College of Rhodesia and Nyasaland in 1960. His Ph.D. was awarded by the University of Cam-

bridge in 1965. He then came to the U.S., where he spent a year at the University of Virginia and three years at Case Western Reserve University. Berridge returned to Cambridge in 1969.

WALTER J. GEHRING ("The Molecular Basis of Development") is professor and chairman of the department of cell biology at the Biocenter of the University of Basel. A native of Zurich, he was educated at the University of Zurich, earning a degree in zoology in 1963 and a doctorate in 1965. While working with fruit flies for his dissertation he became interested in molecular genetics, which has been the focus of his research since then. After spending five years on the faculty at Yale University, he moved back to Switzerland.

Gehring joined the faculty at the Biocenter soon after its founding in 1971.

ALLAN C. WILSON ("The Molecular Basis of Evolution") is professor of biochemistry at the University of California at Berkeley. He got his undergraduate degree in 1955 at the University of Otago in his native New Zealand. He then came to the U.S. for graduate study and received an M.S. at Washington State University (1957) and a Ph.D. from Berkeley (1961). In 1964, after three years of postdoctoral research at Brandeis University, he returned to Berkeley and joined the faculty there. Wilson has also worked at the Weizmann Institute, the University of Nairobi and Harvard University.

# BIBLIOGRAPHIES

*Readers interested in further explanation of the subjects covered by the chapters in this book may find the following lists of publications helpful.*

## THE MOLECULES OF LIFE

PHAGE AND THE ORIGINS OF MOLECULAR BIOLOGY. Edited by Gunther Stent, J. Cairns, James D. Watson et al. Cold Spring Harbor Laboratory, 1966.

MOLECULAR BIOLOGY OF THE CELL. Bruce Alberts, Dennis Bray, Julian Lewis, Martin Raff, Keith Roberts and James D. Watson. Garland Publishing, Inc., 1983.

RECOMBINANT DNA: A SHORT COURSE. James D. Watson, John Tooze and David T. Kurtz. Scientific American Books, Inc., 1983.

## DNA

DNA METHYLATION AND GENE ACTIVITY. Walter Doerfler in *Annual Review of Biochemistry,* Vol. 52, pages 93–124; 1983.

ENHANCER ELEMENTS. George Khoury and Peter Gruss in *Cell,* Vol. 33, No. 2, pages 313–314; June, 1983.

HIGHER ORDER STRUCTURE OF CHROMATIN: ORIENTATION OF NUCLEOSOMES WITHIN THE 30 NM CHROMATIN SOLENOID IS INDEPENDENT OF SPECIES AND SPACER LENGTH. James D. McGhee, Joanne M. Nichol, Gary Felsenfeld and Donald C. Rau in *Cell,* Vol. 33, No. 3, pages 831–841; July, 1983.

THE LOCUS OF SEQUENCE-DIRECTED AND PROTEIN-INDUCED DNA BENDING. Hen-Ming Wu and Donald M. Crothers in *Nature,* Vol. 308, No. 5959, pages 509–513; April 5, 1984.

THE DISTAL TRANSCRIPTION SIGNALS OF THE HERPESVIRUS TK GENE SHARE A COMMON HEXANUCLEOTIDE CONTROL SEQUENCE. Steven L. McKnight, Robert C. Kingsbury, Andrew Spence and Michael Smith in *Cell,* Vol. 37, No. 1, pages 253–262; May, 1984.

INTERACTION OF SPECIFIC NUCLEAR FACTORS WITH THE NUCLEASE-HYPERSENSITIVE REGION OF THE CHICKEN ADULT $\beta$-GLOBIN GENE: NATURE OF THE BINDING DOMAIN. Beverly M. Emerson, Catherine D. Lewis and Gary Felsenfeld in *Cell,* Vol. 41, No. 1, pages 21–30; May, 1985.

## RNA

DO FEATURES OF PRESENT-DAY EUKARYOTIC GENOMES REFLECT ANCIENT SEQUENCES ARRANGEMENTS? James E. Darnell, Jr., in *Evolution Today, Proceedings of the Second International Congress of Systematic and Evolutionary Biology,* edited by G. G. E. Scudder and J. L. Reveal, 1981.

VARIETY IN THE LEVEL OF GENE CONTROL IN EUKARYOTIC CELLS. James E. Darnell, Jr., in *Nature,* Vol. 297, No. 5865, pages 365–371; June 3, 1982.

THE PROCESSING OF RNA. James E. Darnell, Jr., in *Scientific American,* Vol. 249, No. 2, pages 72–82; October, 1983.

RNA SPLICING: THREE THEMES WITH VARIATIONS. Thomas R. Cech in *Cell,* Vol. 34, No. 3, pages 713–716; October, 1983.

SPLICING OF MESSENGER RNA PRECURSORS IS INHIBITED BY ANTISERA TO SMALL NUCLEAR RIBONUCLEOPROTEIN. Richard A. Padgett, Stephen M. Mount, Joan A. Steitz and Phillip A. Sharp in *Cell,* Vol. 35, No. 1, pages 101–107; November, 1983.

## PROTEINS

CATALYSIS IN CHEMISTRY AND ENZYMOLOGY. William P. Jencks. McGraw-Hill Book Company, 1969.

THE ANATOMY AND TAXONOMY OF PROTEIN STRUCTURE. Jane S. Richardson in *Advances in Protein Chemistry,* Vol. 34, pages 167–339; 1981.

SIMILAR AMINO ACID SEQUENCES: CHANCE OR COMMON ANCESTRY? Russell F. Doolittle in *Science,* Vol. 214, No. 4517, pages 149–159; October 9, 1981.

A SIMPLE METHOD FOR DISPLAYING THE HYDROPATHIC CHARACTER OF A PROTEIN. Jack Kyte and Russell F. Doolittle in *Journal of Molecular Biology,* Vol. 157, No. 1, pages 105–132; May 5, 1982.

PRINCIPLES THAT DETERMINE THE STRUCTURE OF PROTEINS. Cyrus Chothia in *Annual Review of Biochemistry,* Vol. 53, pages 537–572; 1984.

A UNIFYING CONCEPT FOR THE AMINO ACID CODE. Rosemarie Swanson in *Bulletin of Mathematical Biology,* Vol. 42, No. 2, pages 187–203; 1984.

## THE MOLECULES OF THE CELL MEMBRANE

MAMMALIAN PLASMA MEMBRANES. Mark S. Bretscher and Martin C. Raff in *Nature,* Vol. 258, No. 5530, pages 43–49; November 6, 1975.

MEMBRANE ASYMMETRY. James E. Rothman and John Lenard in *Science,* Vol. 195, No. 4280, pages 743–753; February 25, 1977.

COATED PITS, COATED VESICLES, AND RECEPTOR-MEDIATED ENDOCYTOSIS. Joseph L. Goldstein, Richard G. W. Anderson and Michael S. Brown in *Nature,* Vol. 279, No. 5715, pages 679–685; June 21, 1979.

MEMBRANE RECYCLING BY COATED VESICLES. Barbara M. F. Pearse and Mark S. Bretscher in *Annual Review of Biochemistry,* Vol. 50, pages 85–101; 1981.

## THE MOLECULES OF THE CELL MATRIX

INTERACTION OF CYTOSKELETAL PROTEINS ON THE HUMAN ERYTHROCYTE MEMBRANE. D. Branton, C. M. Cohen and J. Tyler in *Cell,* Vol. 24, No. 1, pages 24–32; April, 1981.

ORGANIZATION OF THE CYTOPLASM. *Cold Spring Harbor Laboratory Symposia on Quantitative Biology,* Vol. 46, 1982.

TUMOR DIAGNOSIS BY INTERMEDIATE FILAMENT TYPING: A NOVEL TOOL FOR SURGICAL PATHOLOGY. Mary Osborn and Klaus Weber in *Laboratory Investigation,* Vol. 48, No. 4, pages 372–394; April, 1983.

A GUIDED TOUR OF THE LIVING CELL. Christian de Duve. Scientific American Books, Inc., 1984.

MOVEMENT OF MYOSIN COATED BEADS ON ORIENTED FILAMENTS RECONSTITUTED FROM PURIFIED ACTIN. J. A. Spudich, S. J. Kron and M. P. Sheetz in *Nature,* Vol. 315, No. 6020, pages 584–586; June 13, 1985.

## THE MOLECULES OF THE IMMUNE SYSTEM

EVIDENCE FOR SOMATIC REARRANGEMENT OF IMMUNOGLOBULIN GENES CODING FOR VARIABLE AND CONSTANT REGIONS. Nobumichi Hozumi and Susumu Tonegawa in *Proceedings of the National Academy of Sciences of the United States of America,* Vol. 73, No. 10, pages 3628–3632; October, 1976.

SOMATIC GENERATION OF ANTIBODY DIVERSITY. Susumu Tonegawa in *Nature,* Vol. 302, No. 5909, pages 575–581; April 14, 1983.

SEQUENCE RELATIONSHIPS BETWEEN PUTATIVE T-CELL RECEPTOR POLYPEPTIDES AND IMMUNOGLOBULINS. Stephen M. Hedrick, Ellen A. Nielsen, Joshua Kavaler, David I. Cohen and Mark M. Davis in *Nature,* Vol. 308, No. 5955, pages 153–158; March 8, 1984.

A THIRD REARRANGED AND EXPRESSED GENE IN A CLONE OF CYTOTOXIC T LYMPHOCYTES. H. Saito, D. M. Kranz, Y. Takagaki, A. C. Hayday, H. N. Eisen and S. Tonegawa in *Nature,* Vol. 312, No. 5989, pages 36–40; November 1, 1984.

## THE MOLECULAR BASIS OF COMMUNICATION BETWEEN CELLS

DYNAMICS OF STEROID HORMONE RECEPTOR ACTION. Benita Katzenellenbogen in *Annual Review of Physiology,* Vol. 42, pages 17–35; 1980.

BRAIN PEPTIDES: WHAT, WHERE, AND WHY? Dorothy T. Krieger in *Science,* Vol. 222, No. 4627, pages 975–985; December 2, 1983.

STRUCTURE OF MAMMALIAN STEROID RECEPTORS: EVOLVING CONCEPTS AND METHODOLOGICAL DEVELOPMENT. Merry R. Sherman and John Stevens in *Annual Review of Physiology,* Vol. 46, pages 83–105; 1984.

DRUG AND NEUROTRANSMITTER RECEPTORS IN THE BRAIN. Solomon H. Snyder in *Science,* Vol. 224, No. 4644, pages 22–31; April 6, 1984.

THE NATURE AND REGULATION OF THE INSULIN RECEPTOR: STRUCTURE AND FUNCTION. Michael P. Czech in *Annual Review of Physiology,* Vol. 47, pages 357–381; 1985.

## THE MOLECULAR BASIS OF COMMUNICATION WITHIN THE CELL

THE ROLE OF PROTEIN PHOSPHORYLATION IN NEURAL AND HORMONAL CONTROL OF CELLULAR ACTIVITY. Philip Cohen in *Nature,* Vol. 296, No. 5858, pages 613–620; April 15, 1982.

CELLULAR ONCOGENES AND RETROVIRUSES. J. Michael Bishop in *Annual Review of Biochemistry,* Vol. 52, pages 301–354; 1983.

G PROTEINS AND DUAL CONTROL OF ADENYLATE CYCLASE. Alfred G. Gilman in *Cell,* Vol. 36, No. 3, pages 577–579; March, 1984.

THE ROLE OF PROTEIN KINASE C IN CELL SURFACE SIGNAL TRANSDUCTION AND TUMOR PROMOTION. Yasutomi Nishizuka in *Nature,* Vol. 308, No. 5961, pages 693–698; April 19, 1984.

INOSITOL TRIPHOSPHATE, A NOVEL SECOND MESSENGER IN CELLULAR SIGNAL TRANSDUCTION. M. J. Berridge and R. F. Irvine in *Nature,* Vol. 312, No. 5992, pages 315–321; November 22, 1984.

## THE MOLECULAR BASIS OF DEVELOPMENT

A CONSERVED DNA SEQUENCE IN HOMOEOTIC GENES OF THE *DROSOPHILA* ANTENNAPEDIA AND BITHORAX COMPLEXES. W. McGinnis, M. S. Levine, E. Hafen, A. Kuroiwa and W. J. Gehring in *Nature,* Vol. 308, No. 5958, pages 428–433; March 29, 1984.

CLONING OF AN X. LAEVIS GENE EXPRESSED DURING EARLY EMBRYOGENESIS CODING FOR A PEPTIDE REGION HOMOLOGOUS TO DROSOPHILA HOMEOTIC GENES. Andrés E. Carrasco, William McGinnis, Walter J. Gehring and Eddy M. De Robertis in *Cell,* Vol. 37, No. 2, pages 409–414; June, 1984.

A HOMOLOGOUS PROTEIN-CODING SEQUENCE IN DROSOPHILA HOMEOTIC GENES AND ITS CONSERVATION IN OTHER METAZOANS. William McGinnis, Richard L. Garber, Johannes Wirz, Atsushi Kuroiwa and Walter J. Gehring in *Cell,* Vol. 37, No. 2, pages 403–408; June, 1984.

FLY AND FROG HOMOEO DOMAINS SHOW HOMOLOGIES WITH YEAST MATING TYPE REGULATORY PROTEINS. John C. W. Shepherd, William McGinnis, Andrés E. Carrasco, Eddy M. De Robertis and Walter J. Gehring in *Nature,* Vol. 310, No. 5972, pages 70–71; July 5, 1984.

ISOLATION OF A HOMOEO BOX-CONTAINING GENE FROM THE *ENGRAILED* REGION OF *DROSOPHILA* AND THE SPATIAL DISTRIBUTION OF ITS TRANSCRIPTS. Anders Fjose, William J. McGinnis and Walter J. Gehring in *Nature,* Vol. 313, No. 6000, pages 284–289; January 24, 1985.

## THE MOLECULAR BASIS OF EVOLUTION

THE NEUTRAL THEORY OF MOLECULAR EVOLUTION. Motoo Kimura in *Scientific American,* Vol. 241, No. 5, pages 94–104; November, 1979.

BIRDS, BEHAVIOR, AND ANATOMICAL EVOLUTION. Jeff S. Wyles, Joseph G. Kunkel and Allan C. Wilson in *Proceedings of the National Academy of Sciences of the United States of America,* Vol. 80, No. 14, pages 4394–4397; July, 1983.

STOMACH LYSOZYMES OF RUMINANTS, I: DISTRIBUTION AND CATALYTIC PROPERTIES. Deborah E. Dobson, Ellen M. Prager and Allan C. Wilson in *The Journal of Biological Chemistry,* Vol. 259, No. 18, pages 11607–11616; September 25, 1984.

# INDEX

Page numbers in *italics* indicate
   illustrations.

Actin, *60,* 60–65, *62, 63, 66*
   microvilli and, *67*
   proteins which bind to, *65*
Adenine, 4, 14, *16,* 27
Adenosine triphosphate (ATP), *96*
*A* form of QDNA, 14, *18*
Alcohol Dehydrogenase, *38, 42, 43*
Aldosterone, *88*
Allison, James P., 78, 103
Alpha carbon, 41
Alpha helix (of protein molecules), *41,*
   42–43, 52–53, *54, 55*
Altman, Sidney, 35
Amino acids, *40,* 40–42, *41*
   of antibodies, 74
   in enkephalins, *91*
   for homeoboxes, 116–17
   as neurotransmitters, 88
   point mutations and, 120–22
   sequences of, *44,* 44–46
Anderson, Richard G.W., 56
Anion channel, 53
Antibodies, 7, 60, 72–76, *74, 75*
   binding of antigens to, *72*
   genes for, *76,* 76–78, *77*
Antigens, *72,* 73–76, 79
   *T* cells and, 78
Apes, 124, *125*
ATP (adenosine triphosphate), *6,*
   19, 64, *65*
Axel, Richard, 19
Axons, *90*

Bacteria
   ability of mammals to digest, 127
   evolution of, 34–35
   mRNA in, 30
   plasmids in, 5–6
   protein synthesis in, *29*
   regulatory proteins for, 28
Bacteriophages, 5
Bacteriorhodopsin, 53, *55*
Baltimore, David, 36
Bangham, Alec D., 51
Bennett, J. Claude, 76
Berg, Paul, 21
Bernard, Ora, 76, 77

Berridge, Michael J., 96–106
Beta sheet (of protein molecules), *41,*
   42–43
*B* form of DNA, 14, 17, *18*
Binding, by proteins, *38,* 38–39
Birds, evolution of, 128, *129*
Birnstiel, Max, 21
Blood
   cholesterol and ferric ions in, 56
   hormones in, 84, *86*
   *See also* Antibodies
*B* lymphocytes, 72–73, *74,* 77
Brack, Christine, 76, 77
Brains, 127–29, *128*
Branton, Daniel, *66*
Bravo, R., *63*
Brenner, Sydney, 110
Bretscher, Mark S., 50–58
British tits (birds), 128, *129*
Broker, Thomas R., 31
Brown, Michael S., 56
Burgess, Gillian, 103
Burnet, Sir Macfarlane, 72
Burridge, Jane M., *38*
Byers, Timothy J., *66*

*Caenorhabditis elegans,* 109, 110
Calcium, 100–3
Cancer
   intermediate-filament typing and,
     70, *70*
   oncogenes for, 10, 66, 105
   plasma membranes and, 58
CAP (catabolite activator protein), 18
Case, Susan M., 126–127
Catecholamine neurotransmitters, *92*
Cech, Thomas R., 35
Cedar, Howard, 21
Celis, J.E., *63*
Cell membranes, 50–58, *53*
   antibodies bound to, 72
   plasma membranes, *54*
   proteins in, 42
   vesicles in, *50*
Cells
   communications between, 84–93,
     86
   communications within, 96–106, *98*
   of *Drosophila melanogaster,* 112,
     113

early evolution of, *126*
   epithelial, *56*
   evolution of RNA and DNA in, 35–
     36, *36*
   growth of, *104*
   matrix of, 60–70
   plasma membranes of, 58
   specialized, development of, 109
Chambon, Pierre, 20, 21
Cherry, Lorraine M., 126–27
Chien, Y.–H., 79
Chloroplasts, *126*
Cholera, 98–99
Cholesterol, 52, *54,* 55–56
Cholesystokinin, 88
Chow, Louise T., 31, *32*
Chromatin, *22,* 22–23
   structure of, *23*
Chromatosomes, 22
Chromosomes, 110
Chronogenes, 110
*C*–kinase, 103
Clathrin, *50,* 57
Cloning
   of DNA, 6–7
   of genes, 3–4, 7–11, *9, 10*
Codons, 30, 44, *124*
   molecular clock and evolution of,
     122
Cohen, Gerson H., *72*
Communications
   between cells, 84–93, *86*
   within cells, 96–106, *98*
   between neurons, 90
Complex transcription units, 33
Cone, Richard A., 54
Connolly, Michael L., *72*
Corticosterone, *88*
Cortisol, 87, *88*
Crick, Francis, 4, 14
*Cro* protein and repressor, 17, *19*
Crothers, Donald M., 18
Cultural evolution, 128
Cyclic adenosine monophosphate
   (cyclic AMP), *96,* 96–100, *98, 99*
   pathway for, *100–1*
Cyclic guanosine monophosphate
   (cyclic GMP), *98,* 105–6
Cysteine bonding, 42
Cytoplasm, 60, *60*

Cytosine, 4, 14, *16*
Cytoskeleton, *60*, 60–62, *64*, 66
Cytotoxic *T* cells, 80

Darnell, James E., Jr., 26–36
Davidson, Monica, 103
Davies, David R., *72*
Davis, Mark M., 78
Dawson, Rex, 103
Dayhoff, Margaret O., 45–46
De Robertis, Edward, 116
Desmosomes, 68
DNA (deoxyribonucleic acid), *4*
    amino acid sequences determined
        by, 44
    as chromatin, 22–23, *23*
    cloning of, 6–7, *9*, *10*
    discovery of structure of, 4
    of *Drosophila melanogaster,* 115–16
    evolution of, 35–36
    experimental techniques using, 3
    forms of (*A, B,* and *Z*), *18*
    of genes for antibodies, *76*, 76–77,
        *77*
    of genes for *T* cells, 78
    homeobox and, 109, *115*, 117
    of hominoids, *125*
    information transferred to RNA
        from, 26
    inserted into multicellular
        organisms, 8–9
    methylation of, *22*
    molecular clock and evolution of,
        121
    molecular trees (diagrams) of, 123–
        26, *125*
    regulatory molecule for, *14*
    regulatory mutations and, *120*
    repressor proteins and, 17
    restriction-enzyme map of, *8*
    RNA transcription of, 31
    sequencing of, 5
    structure of, *2, 16*
    supercoiling of, 18–19, *20*
    superhelix, *5*
    transcribed to RNA, 27
    Watson-Crick model of, 14
    *See also* Recombinant-DNA
        research
DNA gyrase, 18–19
DNA ligases, 6
Dobzhansky, Theodosius, 44
Doerfler, Walter, 22
Domains (in proteins), 43, 46, 47, *47*
Doolittle, Russell F., 38–47, 105
Dopamine, *92*
Downes, Peter, 103
Dreyer, William J., 76
*Drosophila melanogaster, 109,* 109–
    16, 118
    development of, *111, 112*
    mutations of, *113–17*
    oocyte of, *110*
Drugs, 91–93
Dynein, 67

Eck, Richard, 45–46
Electron microscopy, 2, 62
Elgin, Sarah C.R., 23

Emerson, Beverly M., 23, *23*
Endocrine system, 84, *86, 87*
Endocytosis, 50, *50,* 55–58, *58*
Enhancers, 21–21
Enkephalins, 89–91, *91*
Enzymes, 39–40, 90–91
    binding by, 38–39
    coenzyme-bonding of proteins, *42*
    DNA gyrase, 18–19
    evolution of, 127
    homology in, 45
    lysozyme, *120*
    molecular clock and evolution of,
        121
Epidermal growth factor (EGF), *104,*
    104–5
Epithelial cells, *56*
    cytoskeletons of, 66
    microvilli of, *67*
    protein filaments in, *62*
    proteins from, *63*
Epithelial sheets, 54, 55
*Escherichia coli,* 28
    *lac* operon in, 17
Estradiol, 86–87, *88*
Estrogen, 86
Ethical issues in gene transfers, 11
Eukaryotes
    cell membranes in, 53, *53*
    evolution of, 34–35
    molecular tree of evolution of,
        124–25
    mRNA in, 26, 30, *31*
    protein synthesis in, *29*
    rRNA in, 28
Evans, Ronald, 33
Evolution
    of bacteria, 34–35
    of brain size, *128*
    of early cells, *126*
    gene cloning and, 10–11
    molecular basis of, 120–29, *122–23*
    molecular clock of, *124*
    pressure for, in mammals, *127*
    of proteins, *44,* 44–47, *45*
    of RNA, 35–36, *36*
Exocytosis, *58*
Exons, 32–33, 35, 46–47, 115
Expression vectors, 8

Feldmann, Richard J., *7, 18*
Felsenfeld, Gary, 14–23
Fesenko, Evgenii, 106
Filamentous actin (F-actin), 63, 64, *65*
Finch, John T., 23
Fricker, Lloyd D., 90–91
Frogs, 126

Gallo, Robert C., *8*
Gap junctions, 55, *56*
Garber, Richard, 115
Gastrin, 88
Gehring, Walter J., 109–18
Gelinas, Richard E., 31
Gellert, Martin F., 18, 19
Genes, 3–4
    for antibodies, *76,* 76–78, *77*
    cloning of, 7–10, *9, 10*
    complex transcription units, 33

control over expression of, 19
control over transcription of, 33–
    34, *34–35*
discontinuous, 32
discontinuous, protein evolution
    and, 46–47
of *Drosophila melanogaster, 109,*
    110, 114–16, *115*
duplication of, protein evolution
    via, 44–45
enhancers for, 20–21
homeobox and, 116–18
isolation of, 5
methylation of, *22*
molecular clock for, 120
regulation of expression of, 17
regulatory, 109
regulatory mutations of, 127
regulatory proteins for, 28
role of RNA in expression of, 26
for *T* cells, 78–79
transfers of, 11
for within-cell signal pathways, 105
Genomes, 4, 109
    molecular trees of, 125
Getzoff, Elizabeth D., *6, 72*
Gilbert, A. B., *57*
Gilbert, Walter, 32, 46–47, 76
Gilman, Alfred G., 98
Glenney, John, *66*
Globular actin (G-actin), 63, *65*
Glycolipids, 52
Glycophorin, 53
Goldberg, Michael, 20
Goldstein, Joseph L., 56
Gomperts, Bastion D., 103
Gorter, E., 50
*G* proteins, 96, 99
Grendel, F., 50
Grosschedl, Rudolf, 21
Growth factors, *104,* 104–5
Guanine, 4, 14, *16*
Guanosine triphosphate (GTP), 96, 98
Guy, Robert H., *7*
Gyrase, 18–19

Hafen, Ernst, 118
Haptens, *72*
Haslam, Richard, 103
Henderson, David, *62*
Henderson, Richard, 53, *55*
Heslop, John P., 103
Heuser, John E., *67*
Hogness, David S., 20, 115
Hokfelt, Thomas G.M., 89
Hokin, Lowell E., 102
Hokin, Mabel N., 102
Homeobox, 109, *115,* 116–18
Homeotic genes, 115
Homeotic mutations, 114
Homology, 45, 116
Hood, Leroy E., 76
Hormones, 84–88, *86*
    differential processing by mRNA
        of, *33*
    endocrine system for, *87*
    enkephalins, *91*
    feedback and control mechanisms
        for, *89*

Hormones (continued)
  steroids, 88
  vasopressin, 84
Hozumi, Nobumichi, 76
HTLV-III virus, 8
Humans
  endocrine system of, 87
  evolution of, 128
  gene cloning and study of
    evolution in, 10–11
  menstrual cycle of, 86–87
  molecular trees of evolution of,
    124, 125
  neuropeptide drugs for, 91–93
Hydrogen bonding, 42
Hypothalamus, 86

Immune system, 72, 74
  B lymphocytes, 72–78
  T cells, 78–81
Immunofluorescence microscopy, 62
Immunoglobulins, See Antibodies
Inositol lipids, 100–2, 103–4
Inositol triphosphate (IP₃), 101, 102,
  103–5, 104
Insulin, 6, 88, 106
Intermediate filaments, 60, 62, 62,
  64, 68–70, 69
Introns, 32, 35, 46, 47, 115
Irvine, Robin, 103

Jackson, David P., 23
Jacob, Francois, 28
Jones, Katherine, 21

Kappler, John W., 78, 79
Keratin, 63, 67, 70
Khoury, George, 20
Klessig, Daniel, 31
Klug, Aaron, 23
Kranz, David M., 78, 79
Kunkel, Joseph G., 128
Kuroiwa, Atsushi, 115

Lac operon, 17
Larner, Joseph, 106
L-dopa, 92
Leder, Philip, 76, 105
Levine, Michael, 118
Lewis, Edward B., 114, 118
Ligands, 50
Lipids, 50–53
  inositol, 100–2, 103–4
Low-density lipoprotein (LDL), 58
Lymphocytes, 72–73
Lynch, David R., 90–91
Lysosomes, 57
Lysozyme, 120, 127

McGhee, James D., 23
McGinnis, William J., 116
McKnight, Steven, 19, 21
Magnusson, Staffan, 46
Major histocompatibility complex
  (MHC), 79, 79–81
Mak, Tak W., 78
Mammals, evolution of, 126–29, 126–
  28
Marchesi, Vincent T., 53

Mason, David, 70
Matthews, Brian W., 14, 17, 19
Max, Nelson L., 5
Maxam, Allan, 76
Menstrual cycle, 86–87
Messenger RNA (mRNA), 4, 7, 17, 26,
  28, 31
  differential processing by, 33
  made in eukaryotic cells, 31
  in mutations of Drosophila
    melanogaster, 113
  protein synthesis by, 29
  splicing of, 32, 32–33
  translation into proteins of, 30
Methylation of DNA, 21–22, 22
Mice, 9
Microfilaments, 60, 60–65, 62, 64, 70
Microscopy, electron, 2, 62
Microtrabecular lattice, 69
Microtubules, 60, 60, 62, 64, 66–69,
  68
Microvilli, 66, 67
Mitchell, Robert, 102-3
Mitosis, 60–62
Molecular clock, 120–23, 124
Molecular evolution, 120–29
Molecular paleontology, 45, 46, 46
Molecular trees (diagrams), 123–26,
  125
Monod, Jacques, 28
Morgan, Thomas Hunt, 110
Morphological distance, 126
Murad, Ferid, 106
Mutations
  of antibodies, 77
  in chronogenes, 110
  of Drosophila melanogaster, 113–
    16, 113–17
  neutral, 122–23
  point, accumulation of, 126–27
  point and regulatory, 120
  of proteins, 44–45
Myosin, 64, 65

Neurons, 84, 86
  calcium outside of, 101–2
  communications between, 90
  neurotransmitters released by, 89
  norepinephrine and, 92
Neuropeptides, 89–92
Neurotransmitters, 84, 88–89, 93
  catecholamine, 92
  enkephalins, 91
  vasopressin, 84
Neutral mutations, 120, 122–23, 127
Nicotinamide adenine dinucleotide
  (NAD), 38, 42, 43
Nishizuka, Yasutomi, 103
Norepinephrine, 84, 92, 92–93, 92
Nucleosomes, 22, 22–23
Nucleotides, 14, 16
  of RNA, 26

O'Donnell, T. J., 72
Ohlendorf, Douglas H., 14
Olson, Arthur J., 6, 72, 75
Oncogenes, 10, 66
  within-cell signal pathways and,
    105

Oocytes, 56
  of Drosophila melanogaster, 110
Orgel, Leslie E., 35
Oron, Yoram, 103
Osborn, Mary, 60–70

Padlan, Eduardo A., 72
Paleontology, molecular, 45, 46, 46
Pauling, Linus, 42
Pearse, Barbara M. F., 57
Peptide bonds, 41
Peptides, 84, 87–92
Perry, M. M., 57
Pettijohn, David E., 18
Phages (bacteriophages), 5
Phosphatidylcholine, 50–52, 52
Phospholipids, 50–52, 52, 53, 103
  in plasma membranes, 54
Pituitary, 86, 87
Plasma cells, 73
Plasma membranes, 50, 54, 54–55,
  57–58
  coated vesicles in, 57
  communications within, 96
Plasmids, 5–6, 19
  DNA transfers from, 125
  used in recombinant-DNA
    research, 10
Platelet-derived growth factor
  (PDGF), 104, 105
Point mutations, 120, 126–27
Polymerases, 17, 19, 27
  in eukaryotes, 30
Polypeptides, 41–43
  of antibodies, 75
Poo, Mu-ming, 54
Porter, Keith R., 56
Pozzan, Tullio, 103
Proenkephalin A and B, 90
Progesterone, 88
Prokaryotes
  protein synthesis in, 29
  See also Bacteria
Proteins, 2, 38
  actin-binding, 63–64, 65
  amino acids, 40, 40–41, 41
  bacteriorhodopsin, 55
  binding by, 38
  in cell membranes, 50, 52–53, 57
  of cloned genes, 9
  coenzyme-bonding of, 42
  communications within cells via,
    96
  of cytoskeleton, 60, 60
  enzymes, 39–40
  evolution of, 44, 44–47, 45, 122–23
  in intermediate filaments, 69
  isolation of genes for, 7
  in major histocompatibility
    complex (MHC), 79–80
  point mutations of, 120–22
  polarization and charge of, 41–42
  recognized by immune system, 72
  repressor, 17
  RNA synthesis of, 28–32, 29
  structures of, 42–43
  translation of mRNA into, 30
  X-ray crystallography of, 43–44
Putney, James W., Jr., 103

Rall, Theodore, W., 98
Raulet, David, *80,* 81
Razin, Aharon, 21
Receptor proteins, 38, 50, 96
Recombinant-DNA research, 3
    an antibodies, 76
    gene cloning in, *10*
    products manufactured via, 8
Regulatory mutations, 120, *120,* 127
Regulatory proteins, 28
Reinherz, Ellis L., 78
Repressor proteins, 17
Restriction enzymes, 4
    maps of, *8*
Retinal, *55*
Ribosomal RNA (rRNA), 27–30, 34
    protein synthesis by, *29*
Ribosomes, 26, 28
Rich, Alexander, 14
Ringer, Sydney, 100–1
Rinnerthaler, Gottfried, *60*
RNA (ribonucleic acid), 26–27
    chemical structure of, *28*
    of cloned genes, 9
    control over transcription of genes
        by, *34–35*
    differential processing by, *33*
    differential transcription by, 33–34
    evolution of, 35–36
    messenger, 4
    migration through pores of, *26*
    in production of antibodies, 77
    protein synthesis by, 28–32, *29*
    splicing of, *32,* 32–33
    transcription of, 27–28
    in transcription process, 17
RNA polymerases, 17, 19, 27
Roberts, Richard J., 31
Rodbell, Martin, 98
Rodriguez Boulan, Enrique, 55
Rosenfeld, Michael G., 33
Rossmann, Michael G., 46
Roth, Thomas F., 56

Sabatini, David D., 55
Saito, Haruo, 78, 79
Sarich, Vincent M., 124

Schnaffner, Walter, 20–21
Schulz, Irene, 103
Scott, Matthew P., 116
Sex hormones, *88*
Sex steroids, 86
Sharp, Phillip A., 31
Sherman, William R., 103
Small, J. Victor, *60*
Small nuclear ribonucleoproteins
    (snRNP's), 32
Snyder, Solomon H., 84–93
Sodium channel, 7
Spectrin, *66*
Sphingomyelin, 52
Steitz, Joan A., 32
Steroids, 84–87, *88*
    feedback and control mechanisms
        for, *89*
Streb, Hanspeter, 103
Stress fibers, 60, *62, 64*
Strittmatter, Stephen M., 90–91
Supercoiling of DNA, 18–19, *20*
Sussman, Joel L., *5*
Sutherland, Earl W., 98
SV40 virus, 4, 5, 20
Swanson, Rosemarie M., *40*
Synapses, 84, *90*

Tainer, J.A., *6*
Tainer, John A., *72*
*T*-cell receptors, 72
*T* cells, 78–81, *78–80*
Temin, Howard M., 36
Testosterone, *88*
Thymine, 4, 14, *16*
Tijan, Robert, *21*
*Tk* (thymidine kinase) gene, 19–20, *21*
Tonegawa, Susumu, 72–81
Transcription, 17, 20, 27
    supercoiling and, 19
Transfer RNA (tRNA), 26–30
    protein synthesis by, *29*
Trifonov, Edward N., *5*
Truman, James W., 105
Tryosine, *92*
Tsien, Roger Y., 103
Tubulin, 60, *60, 62–64,* 66–67, *68*

Tumor cells, intermediate-filament
    typing and, 70, *70*

Unwin, Nigel, *26,* 53, 55, *55*
Uracil, 27
Vale, Wylie, 87
Varshavsky, Alexander, 23
Vasopressin, *84*
Vectors, 6, 8
Vergara, Julio, 103
Vesicles, *50,* 51, *53*
    coated, 56, 57, *57, 58*
Vigers, Guy, *50*
Vinculin, *60*
Viruses
    cytotoxic *T* cells to combat, 80
    DNA transfers from, 125
    genes of, 4
    immune response to infection by,
        *74*
    mRNA for, 30
    passing through cell membranes,
        55
    SV40, 20
von Hippel, Peter H., 17

Waterfield, Michael, 105
Watson, James, 4, 14
Weber, Klaus, 60–70
Weigert, Martin, 77
Weinberg, Robert A., 2–11
Weintraub, Harold, 23
Williamson, John R., 103
Wilson, Allan C., 120–29
Woese, Carl R., 34, 35
Wood, William, 23
Worcel, Abraham, 18
Wu, Carl, 23
Wyles, Jeff S., 128

X-ray crystallography, 43–44

Yeast, 10, 117

*Z* form of DNA, 14, 17, *18*